Climate Change Cooperation
in Southern Africa

Climate Change Cooperation
in Southern Africa

edited by
Ian H Rowlands

earthscan
from Routledge

First published in the UK in 1998 by
Earthscan Publications Limited

This edition published 2013 by Earthscan

For a full list of publications please contact:

Earthscan
2 Park Square, Milton Park, Abingdon, Oxon OX14 4RN
Simultaneously published in the USA and Canada by Earthscan
711 Third Avenue, New York, NY 10017

Earthscan is an imprint of the Taylor & Francis Group, an informa business

Copyright © UNEP Collaborating Centre on Energy and Environment, 1998

A catalogue record for this book is available from the British Library

ISBN: 978-1-85383-520-9 (pbk)

Typesetting and page design by PCS Mapping & DTP, Newcastle upon Tyne
Cover design by John Gosling Direct
Cover photograph © Topham Picturepoint

Contents

List of Figures, Tables and Boxes

FIGURES

TABLES

BOXES

List of Contributors

Bothwell Batidzirai is a research fellow with the Southern Centre for Energy and Environment (Harare, Zimbabwe). He has worked previously as an assistant researcher in the Faculty of Engineering at the University of Zimbabwe and as a systems development engineer for a laboratory in Harare. Mr Batidzirai has authored several reports and studies and has presented numerous papers at both local and international meetings. He is a member of the Energy and Climate Change Theme Group of the African Energy Policy Research Network and the National Climate Change Committee. He holds a BSc in Electrical Engineering from the University of Zimbabwe.

Gordon A Mackenzie is a senior energy planner at the UNEP Collaborating Centre on Energy and Environment (Roskilde, Denmark). He has worked previously in the Energy Systems and Environmental Modelling Groups at the Risø National Laboratory (Roskilde, Denmark) and as a deputy director and energy advisor in the Zambian Department of Energy (Lusaka). He is the author or co-author of numerous journal articles and book chapters and co-editor of *Energy Options for Africa: Environmentally Sustainable Alternatives* (Zed Books, 1993). He was one of the initiators of the UNEP Greenhouse Gas Abatement Costing Studies and has worked closely with several African countries on climate change mitigation analysis. Dr Mackenzie holds a PhD in Physics and a BSc in Physics from the University of Edinburgh.

R S (Shakespeare) Maya is the executive director of the Southern Centre for Energy and Environment (Harare, Zimbabwe). He has served as lead author in the IPCC second assessment report's Working Group III and on the IPCC Working Group II's technical paper on regional impacts of climate change. He is the author of several articles in international journals on the economic and political aspects of climate change and the co-editor of *Joint Implementation: Carbon Colonies or Business Opportunities? Weighing the Odds in an Information Vacuum* (Southern Centre, 1996). Dr Maya holds a PhD in energy analysis and policy, a BSc (Hons.) in International Relations and a BSc in Industrial Engineering from the University of Wisconsin–Madison.

Norbert Nziramasanga is the technical director of the Southern Centre for Energy and Environment (Harare, Zimbabwe). He has worked previously as a principal research engineer for the Zimbabwe Electricity Supply

Authority's Corporate Planning Unit and for the City of Harare's Electricity Department. He has undertaken several studies on the power sector, and on energy and resource optimization. He has also conducted numerous training sessions on climate change mitigation. Mr Nziramasanga holds a BSEE in power systems and energy conversion from the University of Michigan.

Ian H Rowlands is an associate professor in the Department of Environment and Resource Studies at the University of Waterloo (Waterloo, Canada). He has worked previously as an energy planner at the UNEP Collaborating Centre on Energy and Environment (Roskilde, Denmark) and a lecturer in International Relations and Development Studies at the London School of Economics and Political Science (London, United Kingdom). Dr Rowlands is the author of *The Politics of Global Atmospheric Change* (Manchester University Press, 1995) and co-editor of *Global Environmental Change and International Relations* (Macmillan, 1992). He has contributed articles on climate change, ozone layer depletion, African energy and development, sustainable development, and business and the environment to many journals and books. He holds a PhD in International Relations from the London School of Economics and Political Science and a BASc in Engineering Science from the University of Toronto.

John K Turkson is an energy economist at the UNEP Collaborating Centre on Energy and Environment (Roskilde, Denmark). He has worked previously as a lecturer in the Department of Planning at the University of Science and Technology in Ghana, and served as a consultant to the World Bank, the United Nations Development Programme (UNDP) and the Ministry of Energy in Ghana. Dr Turkson is the author of several articles on energy planning and policy issues in the energy sector in Ghana. He has also made several presentations at academic conferences on power sector reforms and the transport sector in sub-Saharan Africa. Dr Turkson holds a PhD in Energy Management and Policy from the University of Pennsylvania, an MBA from the Catholic University of Leuven in Belgium and a BA in Economics from the University of Ghana.

Peter Zhou is the director of EECG (Energy, Environment, Computer and Geophysical) Consultants in Gaborone, Botswana. Before founding EECG Consultants, he worked as head of the Geophysics Section in the Government of Zimbabwe, a lecturer in Physics at the Universities of Zimbabwe and Botswana and as a principal consultant with a Botswana water consultancy firm. He has also served as a consultant to various international organizations and individual SADC governments. Dr Zhou has contributed articles to various international journals and book chapters in the fields of mineral and groundwater geophysics, energy, environment and climate change. He is also a principal researcher and theme coordinator of the Energy and Climate Theme Group with the African Energy Policy Research Network. Dr Zhou holds a DPhil from the University of Zimbabwe.

Abbreviations and Acronyms

AfDB	African Development Bank
AFREPREN	African Energy Policy Research Network
AIJ	activities implemented jointly
APEC	Asia–Pacific Economic Cooperation
ASEAN	Association of Southeast Asian Nations
BPC	Botswana Power Corporation
BSD	bulk supply depot
°C	degree Celsius
CAPCO	Central African Power Corporation
CBM	coal-bed methane
CEEEZ	Centre for Energy, Environment, Engineering (Zambia)
CEEST	Centre for Energy, Environment, Science and Technology (Tanzania)
CIDA	Canadian International Development Agency
CNG	compressed natural gas
COMESA	Common Market for Eastern and Southern Africa
DRC	Democratic Republic of the Congo
EC	European Community
ECA	UN Economic Commission for Africa
EdF	Electricité de France
EECG	Energy, Environment, Computer and Geophysical Consultants
EIU	Economist Intelligence Unit
E_t	energy demand in transportation
EU	European Union
FCCC	Framework Convention on Climate Change
FDI	foreign direct investment
GDP	gross domestic product
GEF	Global Environment Facility
GHG	greenhouse gas
GJ	gigajoule
GNP	gross national product
GTZ	Deutsche Gesellschaft für Technische Zusammenarbeit
GW	gigawatt

GWh	gigawatt-hour
HDI	human development index
IDA	International Development Association
IMF	International Monetary Fund
IO	international organization
IPCC	Intergovernmental Panel on Climate Change
IPP	independent power producer
IUCN	World Conservation Union
JI	joint implementation
km	kilometre
kW	kilowatt
kWh	kilowatt-hour
MEPC	Minerals and Energy Policy Centre
MJ	megajoule
MW	megawatt
na	not available
NAFTA	North American Free Trade Agreement
ODA	official development assistance
OECD	Organization for Economic Cooperation and Development
PPP	purchasing power parity
PTA	Preferential Trade Area for Eastern and Southern African States
PV	photovoltaic
SACU	Southern African Customs Union
SADC	Southern African Development Community
SADCC	Southern African Development Coordination Conference
SAD–ELEC	Southern African Development through Electricity
SAPP	Southern African Power Pool
SATCC	Southern Africa Transport and Communications Commission
SCEE	Southern Centre for Energy and Environment (Zimbabwe)
SNEL	Société Nationale d'Electricité (DRC)
t	tonne
TAU	Technical and Administrative Unit
TAZARA	Tanzania–Zambia Railway Authority
TJ	terajoule
TNC	transnational corporation
TWh	terawatt-hour
UN	United Nations
UNCTAD	United Nations Conference on Trade and Development
UNDP	United Nations Development Programme
UNEP	United Nations Environment Programme
UNFCCC	United Nations Framework Convention on Climate Change
WEC	World Energy Council
WHO	World Health Organization

WMO	World Meteorological Organization
WTO	World Trade Organization
ZCCM	Zambian Consolidated Copper Mines
ZESA	Zimbabwe Electricity Supply Authority
ZESCO	Zambia Electricity Supply Corporation Limited

Foreword

Jacqueline Aloisi de Larderel
Director, UNEP Industry and Environment, Paris

The threat of rapid global climate change associated with the build-up of greenhouse gases (GHGs) in the atmosphere is now accepted by a broad spectrum of scientists worldwide. There is also widespread recognition of the need to reduce the risk of climate change by working towards a stabilization of the concentration of atmospheric GHGs. This broad consensus, based on the work of the Intergovernmental Panel on Climate Change (IPCC), is embodied in the United Nations Framework Convention on Climate Change (UNFCCC).

Following the Third Conference of the Parties to the UNFCCC, it is clear that policy discussions are broadening beyond an exclusive focus upon the role of individual nations – hitherto the main unit of analysis in the UNFCCC process – to encompass the role of groups of nations acting together. For example, the Kyoto Protocol highlights the potential for transferring emission-reduction units between different countries. As a consequence, internationally collective action, involving groups of countries, may well contribute to the mitigation of climate change during the early part of the next century. Therefore, this book's publication is particularly timely: it aims to make a first contribution to addressing some of the political, institutional and technical factors that come into play when greenhouse gas emission reduction is considered at the regional level.

UNEP has played a central role in the research related to climate change for many years, through its climate programme based at Nairobi headquarters, and through the formation of the IPCC together with the World Meteorological Organization (WMO). In the analysis of climate change mitigation, UNEP has played a particularly important role by initiating the UNEP Greenhouse Gas Abatement Studies in 1992 through its Collaborating Centre on Energy and Environment at Risø, Denmark. This programme involved both methodological development and country studies. It was an essential step in defining what has become the standard approach to assessing the potential and cost of climate change mitigation measures.

The methodological approach continues to be extended, refined and applied, especially in developing countries and countries with economies in transition. The latest series of country studies and methodological development is being carried out within a project sponsored by the Global

Environment Facility (GEF) through UNEP. In addition to analysis at country level, this project contains a research effort to explore regional mitigation possibilities. Two subregions were selected to be the primary objects of research, although many of the issues and methods are likely to be applicable to any regional grouping. The Andean Group in South America and the countries of the Southern African Development Community (SADC) were chosen for a number of reasons, not least their well-defined political and institutional identities (which include already developed cooperation in the fields of energy and transport – major sectors of interest in climate change mitigation). Another important factor was the existence of collaboration partners within the regions, centres of excellence in research who have already worked together with the UNEP Collaborating Centre on mitigation analysis at national level, and who also have in-depth regional contacts and knowledge.

This book concentrates on the example of southern Africa, and we are grateful for the valuable contributions from the centres of excellence in the region: the Southern Centre for Energy and Environment (SCEE) in Zimbabwe; the Centre for Energy, Environment, Science and Technology (CEEST) in Tanzania; the Energy, Environment, Computer and Geophysical (EECG) Consultants in Botswana; as well as the Centre for Energy, Environment, Engineering (CEEEZ) in Zambia. The former three centres contributed to the chapters, while the latter centre supplied valuable advice and comments at various stages of the process.

Regional cooperation for climate change mitigation, among developing countries, is unlikely to feature as a high priority in the immediate future. The commitments for GHG emission reductions in the Kyoto Protocol are restricted to the industrialized countries, and for good reason. Nevertheless, there are strong arguments for exploring the potential and costs of possible climate change mitigation measures, both at the national level and regionally in the developing world. Measures that reduce GHG emissions, or enhance sinks, in developing countries or regions are, of course, already eligible for support through the Global Environment Facility, provided that the host country or region is a party to the UNFCCC. Moreover, depending upon how the terms of the Clean Development Mechanism are devised, developing country measures, supported by industrialized country partners, could play even more important roles.

UNEP's interest in exploring the issues involved in regional cooperation for climate change mitigation, and the enhancement of local expertise in analysing the possibilities, is in line with the catalytic nature of UNEP's role within the UN system. When regional climate-change mitigation actions eventually do become a reality, as the vast potential indicated by this study indicates they should, the methodological approach for analysing the political, technical and institutional barriers will already be in place, both within the UN system and in the local centres of excellence. These latter centres have the additional role of providing expert and up-to-date knowledge to their host governments. Therefore, the study exemplifies UNEP's role as a catalyst. The present study will also be an important input to UNEP's work on economic instruments.

The specific examples of regional climate change mitigation covered in this book are intended to illustrate the potential and the difficulties. The options covered are far from exhaustive. More detailed work is required, both analytically and organizationally, before these options and similar ones can become 'bankable' options for the international community or other investment agencies. This book indicates that the potential exists. The challenge is to overcome the barriers to exploit the regional possibilities in ways that benefit the member countries of the region as well as the global climate.

Preface

Global climate change has the potential to severely affect all of world society. Shifting climatic zones, higher sea levels and more frequent storms, droughts, floods and heatwaves could generate a range of social, economic and political disruptions. Consequently, it is crucial that various response strategies be identified, analysed and, as appropriate, implemented. In this book, we investigate one set of responses – namely regional mitigation options.

Climate change mitigation refers to the reduction of greenhouse gas emissions or the enhancement of sinks that absorb greenhouse gases. By focusing upon *regional* mitigation *options*, we are interested in those activities that require purposeful coordinated action between or among entities in two or more neighbouring countries that also serve to mitigate global climate change.

Mitigation analysis is, of course, a complex issue; virtually every mitigation option will impact development prospects at local, national, regional and global levels. Key economic sectors – such as energy, agriculture, industry and forestry – all produce greenhouse gases, and all are likely to be affected either directly or indirectly by any mitigation option. Similarly, mitigation options will also have social and political impacts – for example, distributional consequences. As a result, it is crucial that climate change mitigation is investigated within a broader developmental context.

In this book, we examine regional mitigation options, both broadly and specifically. A general framework for analysing regional mitigation options, which has relevance for regions around the world, is developed. Particular proposals for southern Africa are also introduced and scrutinized. More specifically, the investigation we undertake is divided into eight chapters.

In Chapter 1 we justify the particular focus of this book: we show that the question 'How can regional cooperation in southern Africa both promote the mitigation of global climate change and advance the region's development objectives?' is one that is worth investigating. We also introduce the methodology that is followed in the rest of the book. In Chapter 2 we identify a number of factors that could affect the prospects for any kind of regional action – particularly action among neighbouring developing countries. In Chapter 3, attention is focused upon southern Africa more specifically: efforts to promote regional cooperation in this part of the world are reviewed and the potential for further cooperation is surveyed.

A detailed investigation of regional mitigation options involving electricity supply is carried out in the subsequent three chapters. In Chapter 4, the 'baseline' is presented: a scenario for the future development of the region's power sector, in the *absence* of substantial concern about global climate change, is developed. Mitigation options are presented in Chapter 5. These involve greater use of the region's hydropower resources and gas reserves. The abatement potential and associated financial costs are presented for two major mitigation scenarios and a range of mitigation options. In Chapter 6 we consider the broader developmental consequences of the scenarios and options we have developed, as well as the prospects for implementing each. As a result, we are able to evaluate them more comprehensively by considering a wide range of potential advantages and disadvantages.

In Chapter 7 we explore regional options to mitigate global climate change that reside outside of the power sector's supply side. In particular, we investigate the transportation sector and the potential for increasing the use of renewable energy and energy efficiency technologies in southern Africa. For each, we again develop a broad baseline, identify and, where possible, quantify mitigation options, and explore developmental impacts and implementation prospects.

Finally, in Chapter 8 we conclude that regional mitigation options in southern Africa not only exist, but that some offer the potential for considerable carbon dioxide reductions at competitive 'prices'. And though we also conclude that the challenges associated with successful implementation of regional mitigation options are considerable, we argue that their potential – not only in terms of global climate benefits but also in terms of regional developmental gains – certainly encourages their further investigation.

Ian H Rowlands
April 1998

Acknowledgements

This book was completed within a larger project, entitled The Economics of Greenhouse Gas Limitations, which was launched by the United Nations Environment Programme's Collaborating Centre on Energy and Environment in 1996. It comprised eight national and two regional studies in parallel with a methodological development programme. The project was financed by the Global Environment Facility (GEF) through the United Nations Environment Programme (UNEP), and the UNEP Centre was responsible for coordinating the individual studies as well as developing the methodological framework.

Work on this southern African regional study was primarily carried out by researchers at two southern African centres (the Southern Centre for Energy and Environment in Harare, Zimbabwe, and Energy, Environment, Computer and Geophysical Consultants in Gaborone, Botswana) and the UNEP Centre in Roskilde, Denmark. Additional support was provided by colleagues in the region (particularly those at the Centre for Energy, Environment, Science and Technology in Dar es Salaam, Tanzania, and the Centre for Energy, Environment, Engineering in Lusaka, Zambia).

The editor of this book would like to thank all of the contributors to this volume, not only for the quality of their contributions but also for their efforts in commenting on other parts of the work and the development of ideas, more generally. Comments from all participants at a workshop in Harare (in February 1997) served to strengthen the study as well.

The editor would also like to express particular appreciation for the efforts of two people at the UNEP Centre: Gordon Mackenzie, who managed the southern African elements of the larger project for the UNEP Centre, and John Christensen, the head of the UNEP Centre. Gordon was involved in the book's development from start to finish, particularly contributing to all of the methodological developments and helping to facilitate much of the empirical research. John gave his full support and encouragement to the book. Indeed, the editor would like to express his appreciation to all of his former colleagues at the UNEP Centre in Denmark for their support in this endeavour.

A number of individuals – who were not directly involved in the project – looked at draft chapters and discussed ideas and themes related to the book. In particular, the editor would like to thank Ogunlade Davidson, Jørgen Fenhann, Charlie Heaps, Michael Lazarus, Steve Lennon, Robert Redlinger, and Lasse Ringius for their valuable input.

Notwithstanding these comments, responsibility for the individual chapters lies with the authors identified with each chapter. Moreover, although the work was completed largely under UNEP auspices and with GEF funding, the contents do not necessarily represent the views of either the United Nations Environment Programme or the Global Environment Facility.

Chapter 1 | Climate Change Cooperation in the Global Greenhouse

Ian H Rowlands

INTRODUCTION

This chapter has two purposes. Firstly, it aims to justify the particular investigation that is undertaken in the rest of this book: to show that the question 'How can regional cooperation in southern Africa both promote the mitigation of global climate change and advance the region's development objectives?' is one that is worth investigating. Secondly, it is designed to introduce the methodology that will be employed to try to answer this question. To achieve the former, four steps — each consisting of an assertion that must be substantiated — are taken. We take these steps, one at a time, in the first four sections of this chapter. Methodological questions, meanwhile, are addressed in the final section.

In the first section of this chapter, we argue that global climate change is a significant issue. By laying out some of the most important consequences of climate change for world society, we demonstrate that, all else being equal, it would be better for the world not to have enhanced global warming than to have it. Of course, all else will not 'be equal', so it is important to think about how climate change activities and aspirations relate to other goals — whether they are environmental, economic or social, and local, national, regional or global. The significance of global climate change provides the initial justification for this book.

Given that global climate change is worthy of at least some of our attention, the focus logically turns to what should be done. The list of possible actions is virtually endless. After considering some of the most prominent of these, we argue that mitigation options, more specifically those that involve the developing world, should be given high priority.[1] Although these options must not exclude other actions, mitigation strategies need to be a part of society's overall response to global climate change.

The third section directs attention to an even more specific part of our menu of responses. Here, the argument is advanced that research on climate change mitigation in the developing world must include the regional level. Given that extremely attractive possibilities exist at the regional level, at least in theory, analyses of national mitigation options should be complemented by an international focus. Indeed, this addition is justified not only for climate change reasons, but for a whole range of environmental, economic and social reasons. Therefore, the first three steps – taken in the first three sections of this chapter – provide the rationale for studies of regional mitigation options among developing countries.

In the fourth section of this chapter, the book's particular geographical focus is defended and elaborated upon. Recent major transformations in southern Africa – most importantly, the decline of armed conflict and the wave of democratization and economic reform processes – have invigorated discussions about regional cooperation in this part of the world. Moreover, the particular distribution of natural resources and economic activity suggests that coordinated regional action could reduce greenhouse gas emissions or enhance sinks. Relevant climate data, which unveil the sources of greenhouse gases in southern Africa, are also presented; the particular impacts that climate change could have upon the region are enumerated as well.

Having traced out the reasons for investigating the prospects for climate change mitigation in southern Africa, the study's methodology is introduced in the fifth section. Although additional details about the particular approach utilized are presented in subsequent chapters, the final part of this chapter provides the reader with the key elements of the analytical framework employed. Together, then, the sections of this first chapter are designed to justify the particular focus of this book, as well as to provide the basic contextual material into which the subsequent study will be placed.

GLOBAL CLIMATE CHANGE

In 1995, over 2000 of the world's foremost scientists declared that the 'balance of evidence suggests that there is a discernible human influence on global climate' (IPCC, 1996a, p 5). After two to three years of deliberations, the members of the First Working Group of the Intergovernmental Panel on Climate Change (IPCC) concluded that ongoing human actions – in particular, the burning of fossil fuels, landuse changes and agricultural activities – had increased the concentration of so-called greenhouse gases (GHGs) in the atmosphere (IPCC, 1996a, p 3). These GHGs (of which carbon dioxide, methane and nitrous oxides are the most significant anthropogenic ones) 'have led to a positive radiative forcing of climate, tending to warm the surface and to produce other changes of climate' (IPCC, 1996a, p 3). Their report goes on to reveal that global mean-surface air temperature has increased by between 0.3°C and 0.6°C during the past century, and that a further increase of 2°C by the end of the 21st century could well be the result of 'business-as-usual' (IPCC, 1996a, pp 5–6). With assertions like these, the reality of global warming became much more difficult to refute.[2] Figure 1.1 provides a diagram depicting the basic dynamics of global climate change.

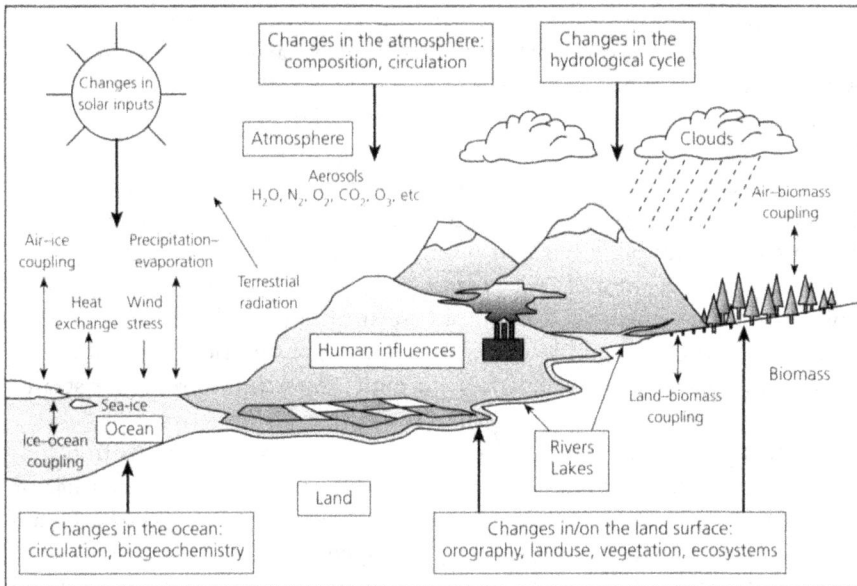

Note: A schematic view of the components of the global climate system (bold), their processes and interactions (thin arrows) and some aspects that may change (bold arrows).
Source: IPCC (1996a) p 55

Figure 1.1 *Global climate change dynamics*

Global climate change may manifest itself in temperature and precipitation changes which, in themselves, may not be such bad things. Indeed, climate has always experienced natural fluctuations over time. What is different about the global climate change issue that is now attracting considerable attention is, firstly, that the change is not natural but rather human induced and, secondly, that the rate of temperature change is set to be unprecedented. The resulting environmental, economic and social consequences of global climate change over the next century could be severe.

The three major environmental consequences of global climate change are, first, a shifting of climatic zones; second, a rise in sea levels; and third, more frequent severe events. Higher temperatures would change the composition and geographic distribution of many ecosystems; some may not be able to provide the same range of goods and services to society, and some may not be able to reach new equilibria for several centuries. Moreover, higher sea levels – rising by perhaps 50 centimetres (IPCC, 1996a, p 6) – would mean that the world's oceans intrude upon precious coastal zones. Finally, more frequent storms, droughts, floods and heatwaves would make the extent of these longer-term variations pale in comparison. Such natural developments would, in turn, have a range of social ramifications. The report of the IPCC's Second Working Group gives examples of the impacts upon human infrastructure, human health and natural resource availability. Many of these impacts would result in significant economic costs and social disruptions (IPCC, 1996b). Therefore, the argument advanced here is that a warmer future for the world would not necessarily be a brighter one. Although there may be unique benefits arising from global climate change, it is widely accepted that, in the absence of other factors, it is more desirable to avoid, than to promote, global climate change.

The choice, however, is not that simple. Although most would presumably prefer not to have global climate change, the calculation is complicated by the fact that the activities that promote climate change themselves generate a range of economic, social and even environmental benefits. The possibility arises, then, that 'the costs are smaller than the benefits' – that is, climate change is an undesirable but relatively insignificant byproduct of largely valuable activities. Following this logic, research into response strategies may well be seen as a waste of scarce resources.

At this point, this 'logic' is rejected. Instead, we adopt the position that it is crucial to think about the appropriate response strategies for either or both of two reasons. Firstly, the possible consequences of climate change are potentially so damaging, that doing nothing could result in disaster – that is, 'the costs could well be much larger than the benefits'. And secondly, the relative costs of taking action may not be as significant as some people believe. In other words, we may be able to 'avoid the costs while still reaping the benefits by different means'.

These observations are not meant to imply that global climate change is the only important challenge facing world society today, and that all possible resources should be mobilized in response. Other severe problems – environmental, social and economic – obviously exist and demand attention. Instead, we simply assert that global climate change has real significance. This provides the motivation for this study.

RESPONSES TO GLOBAL CLIMATE CHANGE

Given that we should be investigating the ways in which the global climate change challenge should be met, the appropriate response must be considered. This particular issue has, not surprisingly, attracted significant attention to date. Generally, much of the effort has been funnelled into studies on the feasibility of either adaptation or mitigation. We briefly consider the former, before concentrating upon the latter.

As noted in the first section of this chapter, the world is already experiencing some climate change. Moreover, given the timelags associated with the impacts of climate change – that is, greenhouse gases emitted at any given time will continue to exercise a positive radiative forcing on climate, thereby increasing temperatures for many years to come – the world is probably committed to more temperature rises. Finally, given the economic and political difficulties associated with realizing a dramatic short-term fall in GHG emissions, human activities are likely to enhance the greenhouse effect, at least during the short term. Together, these observations suggest that climate change is real and will continue to be real.[3]

As a consequence, it is only correct, as part of a comprehensive strategy, to think about the appropriate methods for adapting to global climate change. Fortunately, some of that thinking is already occurring. Much of the 1996 report of the IPCC's Second Working Group, for example, was concerned with adaptability: 'the degree to which adjustments are possible in practices, processes, or structures of systems to projected or actual changes of climate' (IPCC, 1996b, p 5). Moreover, the importance of adaptation has also been recognized in the international negotiations to date: the international treaty guiding action, the United Nations Framework Convention on Climate Change (UNFCCC), commits all parties to 'formulate, implement, publish and regularly update...programmes containing...measures to facilitate adequate adaptation to climate change' (UNFCCC, 1992, Article 4.1(b)). It also pledges them to cooperate 'to prepare for adaptation to the impacts of climate change' (UNFCCC, 1992, Article 4.1(e)), and commits North–South assistance in 'meeting costs of adaptation to [the] adverse effects [of climate change]' for 'the developing country Parties that are particularly vulnerable' (UNFCCC, 1992, Article 4.4).

Another crucial element of the overall response, however, is thinking and action on the issue of climate change mitigation – in other words, the reduction of emissions of greenhouse gases or the enhancement of those sinks that absorb greenhouse gases. This, fortunately, has also received considerable attention to date, among both policy-makers and commentators.

Mitigation is near the top of the list of commitments which all parties to the UNFCCC have accepted: Article 4.1(b) states that 'All Parties...shall formulate, implement, publish and regularly update...programmes containing measures to mitigate climate change by addressing anthropogenic emissions by sources and removals by sinks of all greenhouse gases not controlled by the Montreal Protocol' (UNFCCC, 1992). For the so-called Annex I countries – the then-OECD members, along with many of the countries with economies in transition in Eastern Europe and the Former Soviet Union – the mitigation obligations go further. Article 4.2(a) encourages these countries to ensure that their anthropogenic emissions, and removals by sinks, of greenhouse gases not controlled by the Montreal Protocol are, in the year 2000, no higher than they were in 1990 (UNFCCC, 1992). Given that business-as-usual would have meant an increase in most of these countries, such a commitment implies the need for mitigation actions on their part. Following this, these countries have identified certain policies and measures in their national communications which have been submitted to the UNFCCC Secretariat (National Communications, 1997).

Non-Annex I countries – the developing countries – do not have specific mitigation obligations under the terms of the UNFCCC, nor under the more recent Kyoto Protocol to the Convention (Kyoto Protocol, 1997). This differentiation of commitments between developed and developing countries is accepted by all parties: given that 'the largest share of historical and current global emissions of greenhouse gases has originated in developed countries, that per capita emissions in developing countries are still relatively low and that the share of global emissions originating in developing countries will grow to meet their social and development needs', it is accepted that 'the developed country Parties should take the lead in combating climate change and the adverse effects thereof' (UNFCCC, 1992, Preamble and Article 3.2).

In light of this, one might wonder why thinking about mitigation activities in the developing world is being pursued, particularly given the apparent difficulty that so many Annex I countries are having in meeting their existing mitigation commitments (National Communications, 1997). Indeed, not only might this be a wasteful diversion of research efforts, but it might have much more damaging consequences. More specifically, by focusing attention upon the mitigation potential in the developing world, the fundamental principle of 'common but differentiated responsibility' may be inadvertently (or even knowingly) undermined. When mitigation options in the developing world are highlighted (particularly ones that are cheaper than some in the developed world), is not support being lent to those who claim that Northern action is unjustifiable when 'cheaper options exist in the South (and when their emissions are growing at the rate that they are)'? Of course, this line of argument conveniently ignores issues of justice and fairness which must be central to any global climate change response (Banuri et al, 1996). Still, such interventions may muddy the waters, and the development of fair and efficient global action may consequently be hindered.

Although this potential problem is recognized, it is not accepted as a reason for ignoring research into mitigation strategies in the developing world. Nevertheless, it does remind us of the importance of stressing that the study of mitigation potential in the South is compatible with the execution of mitigation *action* in the North.[4] The 'need for developed countries to take immediate action' has been accepted by all parties to the UNFCCC (UNFCCC, 1992, Preamble). Consequently, any efforts to weaken this message must be quickly rebutted.

Let us reiterate, moreover, that it is not a question of either–or. Instead, thinking about developing country strategies must be part of a comprehensive response to the challenge of global climate change, for a variety of reasons. Firstly, as we have already seen, the terms of the UNFCCC dictate that all parties should develop mitigation programmes. The commitment of non-Annex I parties to this task is contingent upon the provision of 'technical and financial support...in compiling and communicating information' (UNFCCC, 1992, Article 12.7). Support has come from individual OECD countries as well as from international organizations (more about this below). Thus, a fundamental reason for thinking about mitigation in the developing world is that the international community has committed itself to just such a task. Indeed, we have already witnessed the implementation of mitigation projects in the developing world through the Global Environment Facility.[5] Furthermore, particular policies and measures for climate change mitigation have the potential to advance developing countries' own priorities. Not only would less global warming be largely positive for their development prospects – many of the impacts of global climate change would be particularly damaging to the developing world (IPCC, 1996b, passim) – some mitigation activities could also serve to advance developing countries' own environmental, social and economic goals.

A third reason for fostering knowledge about mitigation activities in the developing world is a little more speculative. There is, at present, much talk about joint implementation (JI). This refers to the means where one country can fulfil its obligations by helping to reduce greenhouse gas emissions, enhance carbon sinks or preserve reservoirs in another country. This is advanced in the name of economic efficiency since the marginal costs of reducing emissions or enhancing sinks are different around the world. In the global scheme of things, the argument continues, it makes the greatest sense to pursue the most cost-efficient measures first, irrespective of where they are located. It makes no difference to the climate whether one kilogramme of carbon is emitted in Almonte or Athens, Accra or Ahmadabad – it will contribute to the global phenomenon in the same way and to the same extent.

The debate about JI is heated, to say the least. On one hand, supporters emphasize its economic rationality and argue that JI presents the best chance of securing the involvement of the private sector in the global climate change challenge. Opponents, meanwhile, argue that it is simply a new form of neocolonialism: 'trees for smoke' is one slogan that captures this notion of continued poverty in the South in exchange for continued excesses in the North.[6] Full resolution of this debate is, of course, beyond the scope of this book.[7]

What is extremely relevant for this study, however, is a report on the role that JI is playing, and may continue to play, in the global climate change regime. At the First Conference of the Parties to the UNFCCC, in Berlin in April 1995, parties agreed to establish a pilot phase for JI. During this period, 'activities implemented jointly' (AIJ, as it is being called for this purpose) will proceed among any countries (both North and South) that participate. However, no credits will be forthcoming – that is, no country will be able to meet its year 2000 commitment by activities implemented jointly. Furthermore, parties agreed to 'take into consideration the need for a comprehensive review of the pilot phase in order to take a conclusive decision on the pilot phase and the progression beyond that', before the year 2000 (Report of the Conference of the Parties, 1995).

The Third Conference of the Parties to the UNFCCC, at Kyoto in December 1997, saw the introduction of a 'clean development mechanism' to the discussions (Kyoto Protocol, 1997, Article 12).

> *The purpose of the clean development mechanism shall be to assist Parties not included in Annex I in achieving sustainable development and in contributing to the ultimate objective of the Convention, and to assist Parties included in Annex I in achieving compliance with their quantified emission limitation and reduction commitments.* (Kyoto Protocol, 1997, Article 12.2)

Though the exact terms are still to be negotiated, it appears that this leaves the door for 'effective JI', after the year 2000, clearly open.

What is beyond doubt is that interest in JI and AIJ is increasing. Government officials, businesspeople, representatives of non-governmental organizations (NGOs) and others are examining the issue and considering the role that it might play in meeting the challenge of global climate change. More tangibly, AIJ projects are emerging, with these experiences being fed into the UNFCCC process (through the Conference of the Parties' two subsidiary bodies). Though many of these projects are concentrated in Central America and Eastern Europe, every continent in the developing world has at least one officially recorded AIJ activity (Activities Implemented Jointly, 1996). Indeed, a number of actors who might have instinctively been opposed to JI are now looking at it in more strategic terms, trying to determine how it might best be exploited for the benefit of developing countries (for example, Maya and Gupta, 1996). Nevertheless, just as consideration of mitigation strategies for developing countries does not lessen the imperative for Northern action, exploration of JI possibilities does not imply full endorsement of the scheme. Instead, it is discussed here for strategic reasons; because JI could play an important role in the emerging climate change regime, it is certainly worth exploring the possibilities.

Finally, study into potential mitigation activities in the developing world is justified because it contributes to broader capacity-building on climate change issues in these parts of the globe. The observation that the future of global climate change – in terms of both the natural phenomenon and the social process – will be consequential for the developing world is probably, at best, a gross understatement. Indeed, the fulfilment of the Kyoto Protocol, something that explicitly contains no new commitments for developing countries (Kyoto Protocol, 1997, Article 10), will have considerable implications for developing countries: not only could North–South trade patterns be affected, but precedents for global agreements could (wittingly or unwittingly) be set. If the promotion of mitigation studies can promote 'climate competence' more generally, then that is surely a positive outcome. So, at the end of this second section, we have taken another step: we now have justification – both immediate and longer term – for a study into mitigation proposals in the developing world.

REGIONAL MITIGATION OPTIONS

Thinking about climate change mitigation has been around for as long as people have been thinking that global warming might have adverse impacts. From this, we can estimate that the history of climate change mitigation studies, at least in some form or another, spans over 100 years (Rowlands, 1995, Chapter 3). Concerted international research on the issue, however, has a much shorter history.

As climate change climbed the international agenda during the late 1980s and early 1990s, a number of studies estimating the costs of climate change mitigation were undertaken – most focused upon the OECD countries and, of these, a majority examined the United States (UNEP, 1992, pp 55–64). Nevertheless, some others examined developing countries (for example, IPCC, 1990). However, the quantity of such work increased notably in 1991, at which time the UNEP Governing Council requested the UNEP executive director to initiate a programme of studies to help assess the costs of limiting greenhouse gas emissions. Appointed to coordinate this task was the UNEP Collaborating Centre on Energy and Environment (UNEP, 1992, pp 11–13). Since that time, the issue has been central to the centre's work, coordinating studies with developing country researchers and research institutions. Programmes by the United States government and the German aid agency GTZ (Deutsche Gesellschaft für Technische Zusammenarbeit) have subsequently added additional expertise to the issue of climate change mitigation in developing countries, with much of the focus upon individual countries.[8]

In this study, we complement this work by discussing how regional mitigation options might be conceived, developed and implemented. Our motivation is that regional action will offer better options for meeting the challenges of global climate change, at least in theory.[9] Support for this statement is provided by the observation that including regional action increases the portfolio of available actions, or reduces the cost of existing actions, or permits more equitable outcomes to be realized.[10] Let us present some examples for clarification.

Because any given international region is dealing with entities that are larger than any of its constitutive countries (in terms of population, land area, resources, etc), economies of scale may become critical. For example, when we turn our attention from the individual country to the region as a whole, it could be the case that the benefits arising from the production and distribution of a particular mitigation option (for example, a renewable energy technology) are large enough to encourage an actor to build and sell it. This may be because the inputs are now cheaper (regional suppliers can supply goods at lower costs), because the inputs are now available when they were previously unavailable (enough resources can be captured regionally to afford the initial capital costs), or because production can now be expanded (because of access to a larger regional market). In any case, the result will be the introduction of new mitigation options and/or lower unit costs. This will, in turn, cause the net benefit or cost of that particular mitigation strategy to shift (compared with the calculation performed at the purely national level). So, in the end, we may well be left with a highly attractive regional mitigation strategy, where none was found in the respective national studies.

Furthermore, there may be particular resources which are multinational in character, and which therefore invite multinational management for effective utilization. Such resources – for example, a river, or an offshore natural gas field – may not have been part of a national study because of the perceived difficulties arising from their joint management. However, regional analyses would allow for the introduction of joint resources as potential candidates for the final menu of mitigation options.

Though relatively little work has been completed on regional mitigation options, international society has nevertheless recognized the potential contribution of some kind of regional approach. For one, the UNFCCC specifically refers to regional cooperation as a means of mitigating climate change. Among Article 4's commitments are that: '...All Parties... shall...Formulate, implement, publish and regularly update national and, where appropriate, *regional* programmes' (UNFCCC, 1992, Article 4.1(b), emphasis added).

The UNFCCC also allows the participation of 'regional economic integration organizations' in the international regime. To date, this has basically been a euphemism for the European Community – today the only non-state entity to be a party to the convention. However, there may well be some advantage for a region of the developing world to participate, as a unit, in the global climate regime, rather than as individual countries. References in the UNFCCC to regional economic integration organizations include Article 18 on 'Right to Vote', Article 20 on 'Signature' and Article 22 on 'Ratification, Acceptance, Approval or Accession' (UNFCCC, 1992). The UNFCCC also mentions the possibility of 'joint communications':

> *Any group of Parties may, subject to guidelines adopted by the Conference of the Parties, and to prior notification to the Conference of the Parties, make a joint communication in fulfillment of their obligations under this Article, provided that such a communication includes information on the fulfillment by each of these Parties of its individual obligations under the Convention.*
> (UNFCCC, 1992, Article 12.8)

For a variety of reasons, therefore, there may also be a substantial political advantage to participating in the climate change regime at the regional level. There are also a number of non-climate reasons why it makes sense for developing countries to consider regional cooperative schemes more generally. Given that our particular interest in regional mitigation proposals relates to these, they are worth examining briefly here.

The first justification most often cited is economic. More specifically, it will be in the countries' interests to pursue some kind of regional cooperation (for instance, some form of economic integration), so that they can increase their respective levels of welfare. For example, the benefits arising from economies of scale – in terms of markets, division of labour, specialization in production, and so on – can be reaped. New foreign investment may also be attracted by the arrangements. In total, the proponents argue, the results of greater regional cooperation will be increases in national levels of trade, income and employment. These kinds of arguments, with particular reference to the developing world, are well articulated by Axline (1979, pp 3ff), Blomqvist (1993, pp 52ff), Freer (1996, pp 3–4), Mwase (1995b, p 481) and the South Commission (1990), and by the OECD (1993, pp 25–27) and Robson (1980, p 3 and passim) for the more general case.

There are also, however, political reasons for developing countries to promote cooperative efforts. Many maintain that if developing countries act together, they will have greater bargaining strength in negotiations with external actors – whether they are donors, transnational corporations or international financial institutions. Weeks, for example, argues that 'regional groupings provide perhaps the only viable vehicle by which developing country governments can exert bargaining influence' (1996, p 107; see also Robson, 1980, p 147; and Langhammer and Hiemenz, 1990, p 9). Without such cooperation, external agents may be able to play one country off against another and to reap the gains that otherwise could accrue to the region itself.

Regional cooperation in any one particular area may also spill over into other areas.[11] As a result of an agreement in one area, other policy goals may be furthered (OECD, 1993, p 25; and Langhammer and Hiemenz, 1990, p 10). Vale and Matlosa (1996, p 13) argue that 'Through the process of re-knitting together, a surprising number of everyday problems will be overcome'. Endemic problems of lawlessness, for example, 'may disappear as local communities increasingly take control of dissident groups'. Thompson (1992, p 136), meanwhile, identifies flexibility as a particular benefit of greater regional action: 'Instead of putting up physical or policy barriers, regional decisions find multiple ways of mobilizing resources for solutions to development problems.'

If we can think of these economic and political motivations as being explicitly positive, there are also a couple of justifications which are somewhat more negative. The first relates to the broader phenomenon of regionalization that has been growing during this decade.[12] This leads some to argue that, given greater regionalization in the North – in particular, the widening and deepening of the European Union, as well as the formation of the North American Free Trade Area – developing countries must do the same. Brown, for example, maintains that if developing countries 'do not counter this trend with their own regional integration schemes, they will be effectively shut out of major world markets' (1994, p 2). Mwase also suggests that regionalism must gather steam in developing countries, particularly if subregional exports are 'to be competitive on the international market' (1995b, pp 478–479).

Notwithstanding all of these assertions, it is important to recognize the limitations of regional cooperation as well. The particular arguments above are meant to highlight the *potential* for regional action to generate net benefits – whether they do or not depends, of course, upon the particular kind of regional action undertaken. Indeed, this is true not only for regional cooperation at a very general level, but also for the more specific subset in which we are interested: what we call climate change cooperation.[13] Nevertheless, the main message of this section is that regional mitigation actions are worthy of our attention. This has been justified by reference not only to the developmental gains that can potentially accrue from the mitigation actions themselves, but also from the impetus that they may give to broader efforts at regional cooperation.[14]

SOUTHERN AFRICA AND GLOBAL CLIMATE CHANGE

We now turn our attention to southern Africa in order to justify its selection as a case study, as well as to provide some initial information about its involvement in the global climate change issue.[15] A case study of regional action among developing countries for climate change mitigation could be chosen from various parts of the world. An obvious question is, therefore, 'Why southern Africa?' In response, we offer three reasons.

Firstly, there appear to be interesting parallels among the countries of the region. Though these will be investigated in greater depth in the latter chapters of this book, it is certainly the case that the distribution of natural resources and socioeconomic activity suggests that international cooperation could serve both to mitigate climate change and to advance a broad range of development goals. The potential for electricity interchanges between the region's northern part, which is rich in hydroresources, and the southern part, which has a higher need for such electricity and is limited to its coal deposits, is an example. Indeed, this example forms the basis of much of the substantive investigation in this book (see Chapters 4, 5 and 6). This is not, however, the only candidate, and Chapter 7 reveals how other regional characteristics lead to further regional mitigation options.

In addition to this speculation about regional cooperation, southern Africa also appears to have an encouraging future. Recent transformations, such as the decline of armed conflict and the democratization of many countries in the region, have meant that the prospects for many different kinds of regional cooperation are being enthusiastically investigated in southern Africa. Moreover, it is often forgotten that the region already has a history of successful international relations upon which it can build. Both these past efforts and future potential are examined in greater depth in Chapter 3. For now, it is sufficient to note that because regional cooperation is already on the southern African agenda, the region appears to be a prime candidate for investigating the prospects of climate change cooperation.

Finally, a capacity for climate change investigations is developing in the region. Zimbabwe was one of the first countries to undertake a mitigation study (for example, UNEP, 1993), and similar investigations have also been completed in Botswana, Tanzania and Zambia (UNEP, 1995). Indeed, there is a firm foundation of climate change knowledge upon which this study can be based.

Such characteristics are by no means unique to this region. Nevertheless, they suggest that the basic prerequisites for a regional mitigation study, of the kind undertaken in this book, are in place in southern Africa. Given the novelty of this kind of investigation, and therefore given its exploratory nature, the more advantages that can be secured before commencing, the better. Although not the only candidate, southern Africa certainly appears to be an appropriate region for examination. In order to paint a basic picture of southern Africa's involvement in climate change, information needs to be collected on its contribution to climate change and the potential impacts. Table 1.1 presents information about the emissions of greenhouse gases in each of the countries of the region; the contribution these countries make to global totals are also noted. From these, two things are immediately appar-

Table 1.1 *Estimated emissions of greenhouse gases, southern African countries*

Country	Carbon dioxide emissions from energy-related activities (million tonnes), 1992	Carbon dioxide emissions from landuse change (million tonnes), 1991	Methane from anthropogenic sources (thousand tonnes), 1991
Angola	4.5	16.0	340.0
Botswana	2.2	3.2	110.0
DRC[16]	4.2	280.0	380.0
Lesotho	na	na	44.0
Malawi	0.7	11.0	72.0
Mozambique	1.0	15.0	98.0
Namibia	na	1.8	96.0
South Africa	290.3	14.0	2,400.0
Swaziland	0.3	0.4	25.0
Tanzania	2.1	22.0	760.0
Zambia	2.5	34.0	150.0
Zimbabwe	18.7	5.3	230.0
Southern Africa	326.3	402.7	4,705.0
World	22,339.4	4,100.0	270,000.0
Southern Africa, as percentage of world total	1.5	9.8	1.7

Source: World Resources Institute, 1996

ent: first, South Africa is the region's main contributor (for example, 89 per cent of southern Africa's carbon dioxide from energy-related activities comes from South Africa); and second, even with South Africa, the region's total contribution to global emissions is fairly low, less than 3 per cent of total carbon dioxide emissions.

We now turn to information about the impact that global warming could have upon southern Africa. Though any estimates about regional conse-quences of global climate change are inevitably rough, some nevertheless exist (for example, Magadza, 1994). In addition to the recent report of the Second Working Group of the IPCC – much of which is relevant to south-ern Africa (IPCC, 1996b, particularly pp 438–439) – another study exploring 'some potential impacts and implications [of climate change] in the [southern African] region' has been coordinated by the Climate Research Unit in the United Kingdom (Hulme, 1996). Drawing upon different studies that have been completed, this report considers the impact that three different climate scenarios would have upon a range of different sectors over the next 60 years. Some of the study's main findings are summarized in Box 1.1. Of course, without any discussion of the benefits that climate change activities bring, we are not in a position to assess unequivocally the net consequences of climate change. Nevertheless, many believe that climate effects could be largely negative for southern Africa and, indeed, for Africa more broadly. For example, a recent report by the IPCC concluded that:

Box 1.1 Potential climate change impacts in southern Africa

Over the next 60 years, the possible consequences for southern Africa arising from the 'core scenario for global climate change' include:

- temperature rise of 1.5°C;
- modest drying over large parts of the region;
- decline in grasslands, which are replaced by thorn scrub savannah; expan-sion of dry forest biomes; expansion of desert areas;
- 15 to 20 per cent of large nature reserves and national parks experiencing a change in biome, which has consequences for biological diversity;
- variations in runoff, and greater annual variation in the same;
- crop yield increasing generally, though with some adverse impacts in semi-arid regions;
- changes in the distribution of disease-bearing insects;
- decline in the distribution of ungulate species richness;
- other impacts – for example, sea-level rises – not explicitly considered by the study.

Source: Hulme, 1996

Table 1.2 *Selected economic and social data for southern African countries*

Country	Estimated population (millions), 1994	Real GDP per capita (PPP$US), 1994	Human development index (HDI) value, 1994
Angola	10.5	1,600	0.335
Botswana	1.4	5,367	0.673
DRC	43.9	429	0.381
Lesotho	2.0	1,109	0.457
Malawi	9.6	694	0.320
Mozambique	16.6	986	0.281
Namibia	1.5	4,027	0.570
South Africa	40.6	4,291	0.716
Swaziland	0.8	2,821	0.582
Tanzania	29.2	656	0.357
Zambia	7.9	962	0.369
Zimbabwe	10.9	2,196	0.513

Source: UNDP, 1997

> *The African continent is particularly vulnerable to the impacts of climate change because of factors such as widespread poverty, recurrent droughts, inequitable land distribution and overdependence on rain-fed agriculture. Although adaptation options, including traditional coping strategies, theoretically are available, in practice the human, infrastructural and economic response capacity to effect timely response actions may well be beyond the economic means of some countries.* (Watson et al, 1997)

The potential consequences of global climate change are being viewed with great trepidation by many in the continent.

Finally, to provide additional background information, Table 1.2 presents some basic economic and social data about the 12 countries in southern Africa. These will be complemented by additional data – and a more extensive discussion about the countries' relative similarities and differences – in Chapter 3 of this book. Figure 1.2 presents a map of southern Africa.

METHODOLOGY

Having presented the case for regional mitigation studies in the developing world, the next question is obviously, 'How should they be done?'[17] Given the vast amount of work that has already been undertaken on national mitigation analyses (see the discussion above), it makes good sense to turn our attention there and to review the experience, so that relevant insights can be gained.

The methodology for national mitigation studies is well developed, and the major components are listed in Box 1.2. Though there are obviously

Figure 1.2 *Map of southern Africa*

some variations across individual studies, national mitigation investigations are meant to devote at least some attention to each of these points.

Given that we are interested in the regional level, rather than the national level, what changes must be made to these steps? We consider each of them in turn, in order to determine what meaning they have when translated to this other level.

The first – that is, the overview of the social and economic framework – can be readily examined for the regional level. To a significant extent, this will involve a survey of the region in terms of key statistics (for example, GDP structure, social conditions, energy balance, aggregate greenhouse gas

Box 1.2 Elements of the basic components in national mitigation assessments

1 Overview of national social and economic development framework for climate change mitigation.
2 Baseline scenario projection.
3 Mitigation scenario(s) projection(s), with climate change mitigation and economic costs and benefits across sectors.
4 Macroeconomic assessment.
5 Developmental assessment and analysis of implementation issues.

Source: adapted from UNEP, 1997

inventory, major landuse activities and population). This can be presented in both disaggregated (that is, country level) and aggregated (region level) forms; indeed, some of this material, in the case of southern Africa, has already been presented in this chapter, and further discussion will be undertaken in Chapter 3. What may be different by virtue of the fact that we are concerned with an international region is that there may be wider variations across space; however, that, of course, can also happen within individual countries. Regardless, transferring step 1 to the regional level seems to be perfectly feasible, as well as completely reasonable.

Step 2 consists of constructing the baseline scenario – that is, projections for net greenhouse gas emissions and associated economic information – for the next 20 to 40 years. This would entail the business-as-usual scenario (or the baseline) for both net emissions and costs – in other words, what is projected to happen in the absence of climate change inspired action. To take this to the regional level, we can sum together the various national baselines in order to create a regional baseline. Given that the national baselines are meant to be the best estimate of what business-as-usual would bring at the level of the nation–state, and that together these nation–states constitute the region, it does seem reasonable to proceed in this manner.

Problems, however, may well arise because of the different assumptions that have been made in constructing the national baselines. One country's leaders, for example, may have envisaged a particular level of future trade or a particular trajectory of technological development, while another's may not have foreseen the same. Such incompatibilities do not serve as fundamental challenges to aggregating national projections in this way, and they are certainly surmountable, at least in theory. Nevertheless, they do mean that recalculating some national baselines may be necessary, so that the various projections are compatible.

The third step launches us into a consideration of mitigation strategies. In the national study, this consists of, firstly, identifying the greenhouse gas (GHG) abatement and absorption options. We can think of the same task regionally, with the only qualification being that any regional option must

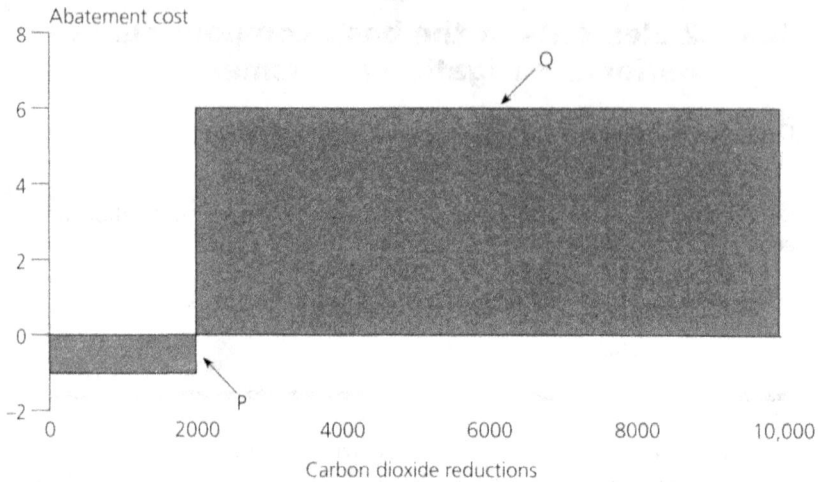

Figure 1.3 *A mitigation scenario that is based solely on regional activity*

require regional cooperation beyond what was anticipated (and built into) the baseline scenario.[18] The kinds of options that would thus qualify as regional have been described by the examples presented earlier in this chapter. The regional analyst will have to think creatively in order to create a menu of options – that is, outcomes which are contingent upon purposeful coordinated action between or among entities in two or more (though not necessarily all) countries. As noted above, this is what we are taking climate change cooperation to be in this book.

Given that we now have this menu of options at the regional level, in addition to the baseline scenario already developed, the task is to compare the two. This allows us to observe the relative gains and losses accompanying regional options – in terms of both climate (net GHG emissions) and economics (money). We will then be able to rank the regional options in terms of their relative 'value'. Just as we do in the national case, we can then think about which options are the most attractive candidates for a mitigation scenario (which will consist of different, though necessarily compatible, options). Therefore, at the end of these procedures, we have a mitigation scenario that is solely made up of regional options: every option is reliant upon some kind of internationally coordinated action.

An example may help to clarify. In Figure 1.3 we have two hypothetical options: 'P' might be the development of a regional market for a particular energy-efficiency device, which not only serves to reduce GHG emissions but can also be secured at a net economic saving. 'Q', meanwhile, may be a large capital project (such as the construction of a hydropower facility), which serves to reduce GHG emissions significantly but at a net economic cost. As usual, potential costs or savings can be found by calculating the areas of the respective boxes.

As a policy tool, Figure 1.3 has, on its own, little value. Why? Simply because it would be very rare for a region to undertake what we have called regional activities in the complete absence of parallel national ones. Instead, what is more useful is a comparison of all options within the region (which is, of course, different from all regional options since the former includes national, as well as regional, options). Therefore, we must think about integrating regional and national mitigation options.

To do this, we take a step back: we need to introduce national mitigation strategies into our deliberations. What is required, first of all, are national mitigation studies for all countries of the region. These, of course, may not always be available. But again, this is not a fundamental challenge to the logic of the analysis; instead, it suggests that many resources may have to be mobilized to realize this requirement.[19] Nevertheless, let us assume that we have comparable national mitigation studies for the countries of the region. It is our assertion that we can take them and simply add them together.[20] We would then be left with a list of nationally based options for the region as a whole. Figure 1.4 presents an example of two countries' mitigation curves, along with their summation. The sizes of the blocks are not affected by the summation since we have assumed that national options develop independently of each other.

In words, Figure 1.4 depicts the following situation: if a region wants to mitigate climate change in the most cost-effective manner, but without any interaction among its constituent countries, then the options identified in the bottom curve would be selected. (As is the procedure for individual countries, selection is done by working from left to right along the curve.) This strategy would be more efficient, for the region as a whole, than assigning arbitrary mitigation targets (for example, equal percentage abatement) to each of the two countries individually. Again, however, this particular figure, on its own, may have little use in reality.[21]

The value emerges when we integrate these two scenarios − that is, we bring together Figure 1.3 (the scenario that is based solely on regional activity) and Figure 1.4 (the scenario by summing national scenarios). This, however, is not simply a case of summing the two since there may well be incompatibilities. For example, part of a national scenario may be fuel substitution in electricity generation. This plan would, of course, be made redundant if part of the regional scenario were to generate all of the region's electricity by a new hydropower facility. Consequently, introducing the regional scenario will cause some elements of the national scenario to disappear or drop out (as in the case of the example above). Other elements of the national scenarios − those that will not be affected by the new regional cooperation envisaged in the regional scenario − will remain the same (for example, an agreement to liberalize trade in energy-efficiency devices may have virtually no impact upon a particular transportation related option). Still others may be affected but not eliminated (the development of a regional market would affect the costs of the mitigation options that were

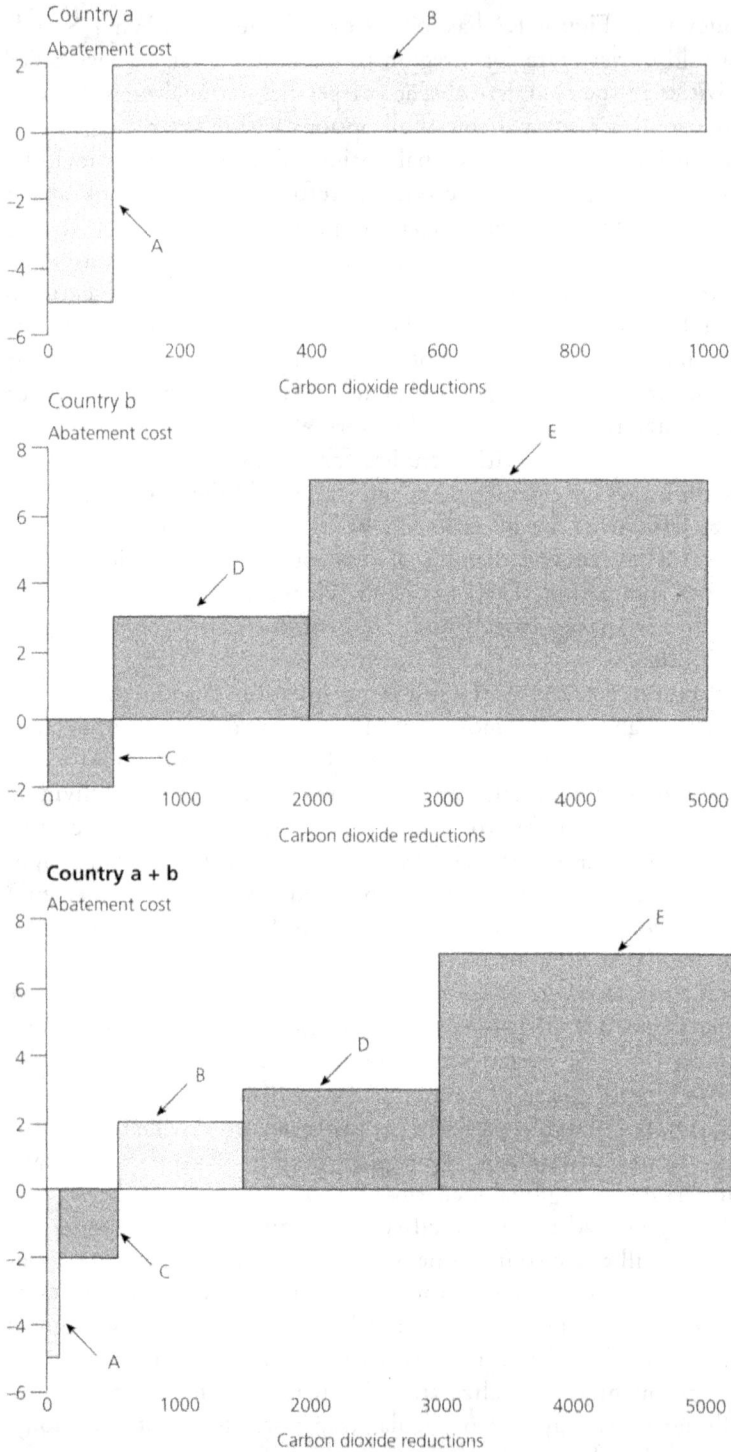

Figure 1.4 *A mitigation scenario by summing national scenarios*

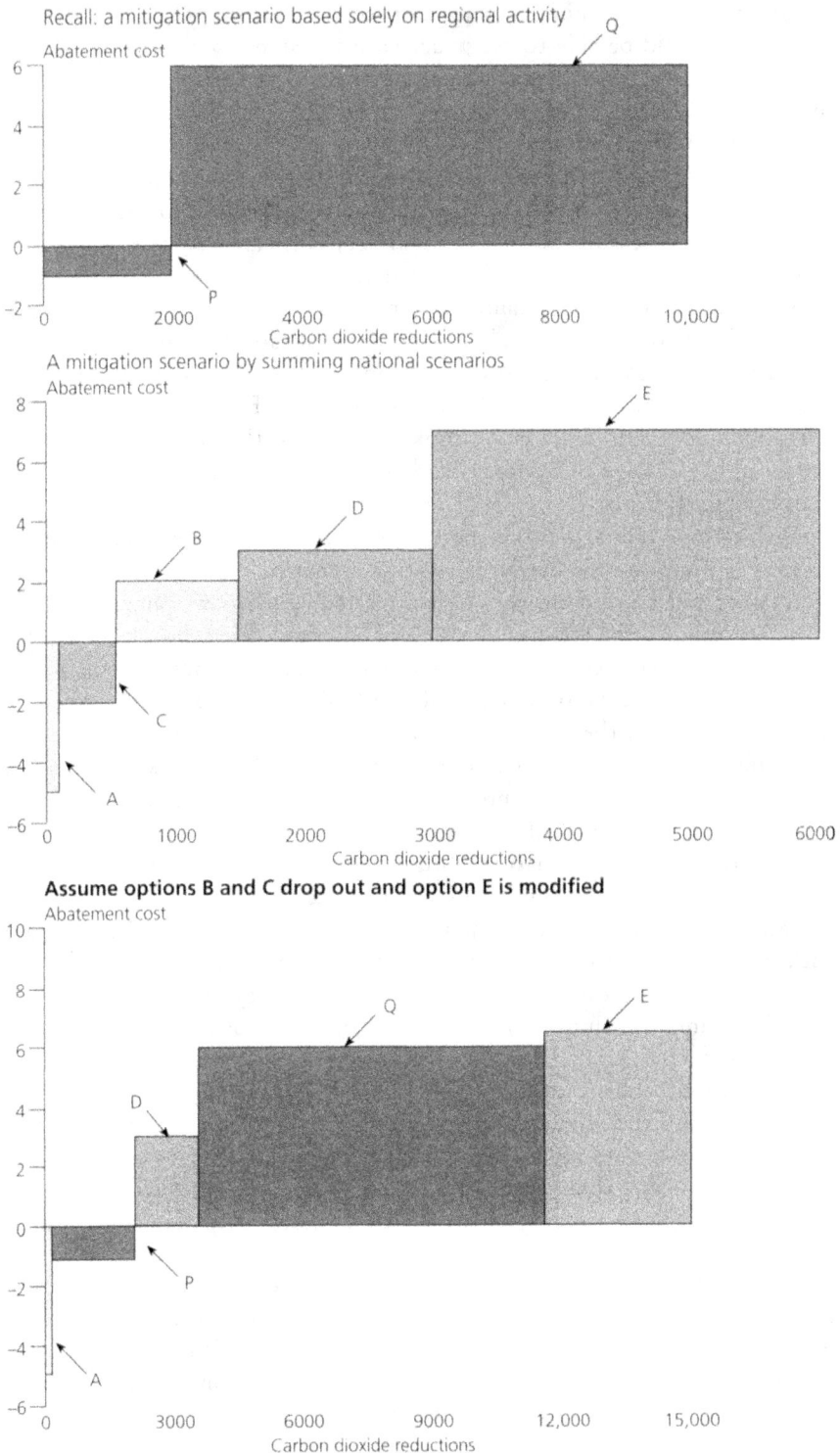

Figure 1.5 *A regional mitigation scenario*

initially considered within the context of solely a national market). Given all of this, we would be able to construct an integrated regional scenario (see Figure 1.5). Note that in this example, options B and C have dropped out, and the shape of option E has changed, all because of the interaction among national and regional options.

We have, therefore, a regional mitigation scenario made up of both national and regional options. Given this, we should check to ensure that it is better than the sum of the national scenarios already developed. For instance, while it is certainly the case that a better regional solution will always exist, there is no guarantee that the analyst has necessarily found it. In the desire for 'regionality', a poor regional solution may have been selected. The message is simply to remember that regional cooperation is not necessarily good or better. So we are left with Figure 1.5 – a regional mitigation scenario. This curve looks like those that have already been produced for a range of national studies. What it will clearly show is that the best portfolio of actions will involve a combination of national and regional actions. Thus, it also demonstrates that the introduction of regional options can improve the overall mitigation scenario.

Having constructed our regional mitigation scenario, we can now analyse it by returning to the steps of the national methodology (Box 1.2). We see that the fourth step consists of undertaking a macroeconomic assessment. How can we move this to the regional level? Given the paucity of macroeconomic modelling at the national level in many developing countries, this will inevitably be quite a challenge at the regional level. Nevertheless, once again, there is nothing in theory to prevent this – we will begin to deal fully with the crucial 'in practice' challenge below.

The final step in the national studies involves a developmental assessment and an analysis of implementation issues. The former appraises the scenario on a range of other yardsticks to see how compatible the plans are with other aspirations. We can, and of course should, do the same for the regional scenarios, measuring them against a variety of developmental indicators. These might include social issues such as equity or local environmental indicators.[22]

The difference between the national studies and the regional studies is at its greatest with the implementation issues. Though certainly demanding at the national level, they are particularly challenging at the regional level. This stems from the fact that national-level studies are concerned with individual countries, which are sovereign within their own territory. Regional studies, alternatively, bring together a number of countries, each of which maintains its sovereignty when it acts in the international system. Because of the nature of the interactions amongst these different units (that is, states), a whole new set of challenges to implementation arise. Indeed, this is where the most methodological value is added in this study of regional mitigation prospects: we explicitly (theoretically in Chapter 2, and empirically in Chapters 3, 6 and parts of 7) investigate the extent to which political and institutional aspects

---

Box 1.3 Using the national study to guide the regional study

National	Regional
National	*Regional*
1 Gather relevant data.	1 Gather relevant data.
2 Construct the reference or baseline scenario.	2 Construct the reference or baseline scenario.
3 Develop the mitigation scenario(s): • identify the options; • assess the net costs and GHG reduction and absorption potential; • create a mitigation scenario.	3 Develop the mitigation scenario(s): • identify the options; • assess the net costs and GHG reduction/absorption potential; • create a mitigation scenario that is based solely on regional activity; • create a mitigation scenario by summing national scenarios; • integrate the last two steps to create a regional mitigation scenario.
4 Macroeconomic assessment.	4 Macroeconomic assessment.
5 Developmental assessment and analysis of implementation issues.	5 Developmental assessment and analysis of implementation issues.

(particularly those that arise by virtue of the need for international interaction) affect the prospects for successful climate change cooperation.

In summary, Box 1.3 shows how the national study methodology has been used as a guide to generate a methodology for a regional study. Remember, of course, that the whole process is iterative – as is the case in the national level study. The introduction of different regional options will generate different mitigation scenarios that are based solely on regional activity. These will, in turn, lead on to different regional mitigation scenarios.

What we must now consider is the feasibility of the steps listed in the right-hand column of Box 1.3. As hinted at already in this section, it might well prove difficult to go from theory (the discussion above) to practice (actually undertake a regional mitigation investigation, as is the stated task of this book). For steps 1 and 2 – given the paucity of information in many parts of the developing world – data collection could be a particular challenge. To overcome this, we may well have to make assumptions. These might involve estimating values for particular sectors or deciding, because a region is dominated by some subset of its countries, that the analysis will proceed solely with data from a limited number of countries.

More challenges arise when we turn to the third step. In addition to the problems already foreseen (in particular, trying to bring together the different national mitigation scenarios), integrating the regional and national options (in particular, determining how they impact each other) will be particularly difficult. This is not only because of data limitations, but because of a more general lack of understanding about how different sectors among different countries relate to each other.

As a result, the construction of a curve like that in Figure 1.5 may well prove too difficult to complete. Given that regional mitigation studies represent new terrain, we should not feel too disappointed by accepting this and consequently revising our expectations. What would prove useful, to both policy-makers and theoreticians, would be clear information about particular regional options – that is, full elaboration on what is technically required, an estimation of the amount of GHG abated or absorbed and, finally, an approximation of the cost of successfully executing such an option (which could then be translated into units of money per unit of GHG). These quantitative data could subsequently be followed by a qualitative discussion of how these regional options compare to the already-calculated national options. Part of this discussion would involve what we have presently labelled the fourth step. Given the paucity of data and models to calculate macroeconomic impacts at the national level, we should be fully content with a more modest ad hoc elaboration of macroeconomic consequences at the regional level.

Approaching the fifth step, therefore, our analysis would have thus far produced a regional option, complete with information about its mitigation potential and associated costs or savings. It would also have produced some preliminary impressions of the way in which the option will impact upon other relevant activities in the region. This would allow us to compare it with the list of options that are already present in the region, and which do not depend upon the kind of regional cooperation that we have envisaged. If we are still interested in the regional option, then we should perform the developmental assessment and analysis of implementation issues. Eventually, therefore, we should have a relatively good idea about the consequences of a particular regional policy, as well as some information about its broader (non-climate) impacts and prospects for successful realization. What we will not have – and what might have been expected were the regional analysis to deliver the same kinds of results as the national studies – is more specific information (that is, numerical data) about the likely impacts of the regional mitigation options upon all sectors of the region's economies. Moreover, because we will not necessarily have examined all of the regional options but, instead, only a few selected ones, we will not be able to make explicit policy recommendations. The research aim of our task – to develop thinking about regional mitigation strategies, while nevertheless applying our ideas to some case studies – remains foremost in our minds.

Nevertheless, the various steps that we have laid out will allow us to derive and to analyse a limited range of regional mitigation options. This analysis will be initially undertaken (Chapter 5 and parts of Chapter 7) on climate and economic criteria, which will generate options that are the most attractive in terms of costs per unit GHG abated and absorbed. Following this (Chapter 6 and other parts of Chapter 7) is a developmental assessment and analysis of implementation issues of the same set of options, which will generate more information about their desirability as well as their prospects for successful implementation. This may well modify the relative attractiveness of the different proposals.

By pursuing the investigation in this manner, some conclusions will be advanced at the end of the book. They will identify which, if any, of the regional mitigation proposals appear to warrant further study. Reflection upon the study will suggest additional proposals vis-à-vis the methodology for regional mitigation studies. In particular, we will be able to decide whether our methodological ambitions were appropriate or not, and how they might be refined for future studies.

SUMMARY

This chapter had two aims: to justify the particular investigation that is undertaken in the rest of this book, and to lay out the methodology that will be used. To fulfil the former, a number of small investigations were taken in the first four sections of the chapter: the significance of global climate change was demonstrated, and the value of mitigation studies in developing countries – and, more particularly, a regional study in southern Africa – was illustrated. To fulfil the chapter's second aim, an 'ideal' methodology was developed by using the existing steps for the national studies as a guide; this was subsequently tempered in the face of significant challenges.

We hope that the reader now has a good idea of the particular context, both empirical and theoretical, within which this study is located. It is now time to move on to discuss the most valuable methodological contribution of this study: the development of a framework for investigating the prospects for successfully implementing regional mitigation strategies. This is the purpose of Chapter 2.

ENDNOTES

1   Mitigation options consist of policies and measures that are designed to reduce emissions of greenhouse gases or to enhance sinks of the same.
2   Some, of course, continue to refute it. For sceptical views on the science of global climate change, see, for example, George C Marshall Institute (1998). These kinds of studies notwithstanding, the IPCC report claimed that it represented a significant share of scientific opinion (eg, Bolin and Houghton, 1995, p 176).
3   The IPCC argued that the temperature change we have already experienced is unlikely to be entirely natural in origin (IPCC, 1996a, p 5).
4   While discussion of 'mitigation' in Annex I countries usually implies stabilization or reduction of net emissions, the same discussion for developing countries usually implies reducing net emissions from what they would otherwise have been; this usually still results in an absolute increase over time. It is in this spirit that we use the term mitigation.

5   Moreover, many accept that a global cap on net emissions will have to considered at some point. This implies that all entities will have to consider, at a minimum, stabilization measures.

6   For one sceptical view of JI, see Harvey and Bush, 1997.

7   See, for example, the range of references gathered in the issues of the quarterly publication *Joint Implementation News*, published by the Foundation JIN, Groningen, The Netherlands.

8   For the broader policy context, see UNEP, 1997, Chapter 2. For a review of mitigation studies, see Haites and Rose, 1996.

9   In theory, there is the limiting case, which suggests that the region would be able to do no better than the states when acting on their own. However, the chances of us encountering this limiting case are virtually nil, in reality.

10   'Under uncertainty, a portfolio of measures will, on average, yield a better outcome than any individual action' (Arrow, Parikh and Pillet, 1996, p 72).

11   The language of functionalism is being used purposefully here. The way in which functionalist theory helps us to understand regional cooperative efforts in the developing world will be investigated in Chapter 2.

12   This discussion will be elaborated upon at the beginning of Chapter 2.

13   For the purposes of this study, we take climate change cooperation to be those mitigation options and scenarios which require purposeful coordinated action between or among entities in two or more countries of the region.

14   Moreover, we recognise that more general (that is, non-climate) movement towards regionalization might also encourage climate change cooperation.

15   In this study, we take southern Africa to consist of those 12 mainland African countries south of (and including) the Democratic Republic of the Congo and Tanzania. See the map in Figure 1.2 for further details.

16   The country did not become formally known as the Democratic Republic of the Congo (DRC) until 1997; before that time, it was known as Zaïre.

17   Though much of this book focuses upon southern Africa, we strive to keep the discussion in this section at a general level.

18   It could well be confusing if some elements of a regional strategy have already been included in the national scenario and thus already form part of the regional baseline. We will be attentive to this possibility.

19   And, just as was the case in the aggregation of the baseline scenarios, we will have to be aware of the extent to which the national scenarios are compatible.

20   The rationale is the same as was advanced for the summation of the national baselines.

21   If there is going to be sufficient regional cooperation to agree to such a strategy – that is, to choose the cheapest option, regardless of where it is located – then there would most likely be sufficient regional cooperation to generate new and potentially better regional options!

22   It is difficult to say anything more specific about the developmental assessment; it will be determined by the particular region selected.

Chapter 2 | Explaining Regional Cooperation

Ian H Rowlands

INTRODUCTION

Any kind of arrangement that is dependent upon coordinated action between at least two players always runs the risk of breaking down. One side may not be able to keep up its end of the bargain. Alternatively, it may not be a case of one side being unable to cooperate but, instead, of being unwilling to cooperate: the temptations offered by a strategy of non-cooperation may prove irresistible. Such a possibility is endemic to every cooperative arrangement; as a result, the risk of cooperative arrangements breaking down, or failing to materialize in the first place, is always present (for example, Hardin, 1968).

International cooperative arrangements are, of course, by no means exempt from this state of affairs. Indeed, given the fact that there are much weaker enforcement mechanisms at the international level, it is often the case that international arrangements are that much more difficult to realize. We should not, therefore, expect any kind of regional cooperation – climate change cooperation included – to come easily.[1] Even if schemes that appear to have universal benefits are envisaged, a variety of factors may preclude their effective fulfilment. Given this, a question that immediately arises is: 'How might regional cooperation, particularly among developing countries, be encouraged and sustained?'[2] The primary purpose of this chapter is to present some initial ideas in response.

THE EXPERIENCE THUS FAR

Before launching into this inquiry with both proverbial feet, it is useful to review the past experience of regional cooperation among developing countries. Given our interests in this book, regional cooperation could conceivably come in many varied forms. It includes, for example, the four forms of integration that are usually the primary concern of economists. Moving from the less to the more substantial, they are: free trade areas, customs unions, common markets and economic unions (Balassa, 1961, quoted in Vaitsos, 1978, pp 720–721). But regional cooperation also includes other actions that are aimed at enhancing the common interests of the regional participants, though do not qualify as any kind of economic integration. This might involve various kinds of political or cultural arrangements – for example, a treaty concerning mutual defence or an agreement to encourage cross-border sporting competitions. As a consequence, an exhaustive survey of regional cooperation, even restricted to the developing world, would embrace a huge range of activities.

Given the interest accorded to economic integration in both the practice and theory of regional cooperation, it makes sense to begin our brief overview there. Among developing countries, a number of analysts argue that there have been two major phases of regional economic integration (for example, de Melo and Panagariya, 1993; Mytelka, 1994; and Park, 1995). The first, from the late 1950s to the 1970s, was motivated by a desire to support import-substitution development strategies (Park, 1995, p 25). Economic cooperation among developing countries was seen to provide sufficiently large markets for indigenously produced goods. Thus, this first wave was primarily inward looking, seeking to delink developing countries' economies from the industrialized Northern ones. Most of these kinds of arrangements took place in Latin America and Africa, with the Andean Pact (established in 1969) and the Economic Community of West African States (established in 1975) just two such examples.

The post-1980 period has seen a second wave of initiatives for regional economic integration within the developing world. This 'new regionalism', as many call it (for example, de Melo and Panagariya, 1993), is much more outwardly focused, concentrating upon export-led strategies for economic development. It is still intended to advance a regional orientation, but does so without delinking from the broader multilateral (global) environment. Examples of such regional initiatives include the Southern Common Market (MERCOSUR, signed in 1991) and the ASEAN Free Trade Agreement, signed by the six members of the Association of Southeast Asian Nations (ASEAN) in 1992, (Park, 1995, p 35). Moreover, a number of arrangements have involved a combination of Northern and Southern states – for example, the Asia–Pacific Economic Cooperation (APEC) and the North American Free Trade Agreement (NAFTA).

Recognition of these two periods leads to two important conclusions. The first, and the most obvious, is that there have been efforts at regional integration among developing countries in the past. The second, meanwhile, is that the variation across (and within) the two periods shows that individual arrangements have had different goals and strategies. The former implies that the past may teach us something, while the latter warns us that any lessons from the past will by no means be universally applicable.

Beyond integration schemes, there have been a range of politically motivated regional arrangements. ASEAN's establishment in 1967, for example, was primarily for security reasons. There have also been numerous instances of cooperation without organization – that is, arrangements that have not had formal bodies associated with them. This would be an example of international cooperation in its broadest sense. Following from this, we must remember that climate change cooperation could take any (or all) of these forms.

Recognizing that there has been significant activity regarding regional cooperation among developing countries, a logical next step is to assess this record: were the articulated goals achieved? Were any benefits that could not have been achieved by merely unilateral activity realized? Alternatively, did the product result in more harm than good? There have been a number of specific investigations which have concluded that individual arrangements were positive. ASEAN is probably the one that is most often lauded. Blomqvist, for example, asks in the title of a recent article whether it can be 'a model for third world regional economic cooperation' (1993). Park also highlights the success of ASEAN, arguing that it has been so successful because it has not restricted itself to integration, but instead looked at 'political and economic cooperation for development' more broadly (1995, p 31).[3] In addition to ASEAN, there have no doubt been other particular arrangements that have also generated more benefits than would otherwise have arisen.

Against these isolated incidents of success, however, a number of commentators suggest that the disappointments are much more prominent. Surveying regional cooperation efforts more generally, Vaitsos (1978, p 719) reported that most were in 'various degrees of serious crisis'; Mytelka (1994, p 25) writes of the 'failure of regional integration schemes'; Kisanga (1991, p 13) argues that their performance 'has not been successful'; and Zormelo (1995, p 4) comments on the 'low level of achievement'. More systematically, an OECD study, after examining 12 regional trading arrangements, found that the experiences were 'rather disappointing' (OECD, 1993, p 10). For their part, Langhammer and Hiemenz (1990, p 59) examine regional integration in three major areas of the developing world and conclude that 'the expected benefits ... have not materialised'.

Given the instances of both success and failure in regional cooperation among developing countries, one question is: 'What might bring about either the former or the latter?' Since this is the central concern of this chapter, we turn to it now.

OFFERING EXPLANATIONS

To develop some ideas about what might lead to either success or failure in regional cooperation efforts in developing countries, we seek direction from a variety of sources. An obvious place to look is the literature that has addressed this specific question. Though Thompson (1992, pp 135–136) argues that the 'development of new theories of regional cooperation, based on concrete experiences of developing countries, is in its infancy', there are, nevertheless, still some studies that yield insights. Secondly, we look to those who have investigated the experiences of specific efforts at regional cooperation, not only among developing countries, but more generally as well. Finally, we are open to any other contributions that might assist us in answering the central research question that has been posed in this chapter. After contemplation of these various sources, we find that the question can be usefully answered, at least potentially or partially, by attention to ten factors. The subsequent sections of this chapter consider these.

Sovereignty

In many cases, the benefits stemming from regional cooperation will only arise once some degree of decision-making power has been relinquished to a supranational body or, alternatively, pooled in a regional effort. This perceived ceding of sovereignty has often been one of the main barriers to regional arrangements, not only in the developing world but across the globe.

Many, however, argue that it has been a particularly significant impediment in the developing world. Because many developing countries have only recently gained their independence, they instinctively resist efforts to interfere with the exercise of that independence. A cooperative effort could conceivably retard the development of a unique and uniform country identity by diluting various national traits in a regional cocktail. Blomqvist (1993, p 54), for example, argues that many developing countries 'are still in the process of finding their national identity; therefore it is common that [they] will guard strongly against any sacrifice of their newly won sovereignty, and borders will consequently gain in importance'.

Numerous studies confirm that this has been a major barrier to building regional arrangements in the developing world. A 1993 OECD study (p 66) concluded that the 'surrendering of (some) sovereignty over economic development was a sacrifice [many developing countries] were not prepared to make'. Moving from the systematic investigation to the somewhat more anecdotal, Johnson (1991, p 4) tells his readers that 'any one who has sat through these discussions [on potential regional arrangements] cannot fail to see that there is an underlying reluctance by leaders to lose control over economic policies that directly affect their own nationals'. For the purposes of this study, we are left to believe that concerns about the relinquishing of sovereignty could be an impediment to regional cooperation.[4]

Elite Interests

It is also useful to concentrate upon the concept of interests. In other words, it is helpful to focus attention upon the ways in which the candidates, in a potential cooperative arrangement, view the benefits and costs associated with participation.

However, why should interests be a factor? If a particular proposal for regional cooperation is being mooted, then it must be because it is a good thing, one that will bring benefits to all involved. If it were not, then why would the particular scheme even be considered by the potential participants?[5] Consequently, why worry about interests; if the only impediment were interests, then there would be no obstacle to cooperation at all: rather, all involved would recognize the inevitable benefits and proceed accordingly.

This is by no means always the case, because individuals may not see things in the same way that an objective analyst does. In other words, although the observer may conclude that there will be certain benefits for all participants in a particular cooperative scheme, this conclusion may be based upon certain assumptions – in particular, the analyst's own understanding of what is in the actor's interests, or what is rational.

Brown is one who concentrates upon this theme. She maintains that there are a number of inherent problems with any kind of 'rational actor model', because there is no single rational choice for a particular actor (Brown, 1994, p 14). Instead, because actors shape decisions within their own distinctive cognitive frameworks, particular factors will be given different weightings by different actors. When applied to a study of regional cooperation among developing countries, Brown (1994, p 3) argues that we should look to developing countries' elites, and recognize that they 'frame decisions regarding regional cooperation in terms of short-term, national, and personal loss-avoidance'.[6] In other words, though a particular economic analysis may yield one calculation of relative costs and benefits, perhaps aggregated at the regional level, a more cognitive analysis would involve a different kind of calculation, placing greater emphasis upon more immediate, subregional consequences.

While focusing upon the interests of developing countries' elites, a few points merit elaboration. Firstly, this emphasis upon the short term is important. Given the demands of elections, or the need to ensure political survival more generally, leaders may well opt for the certain (though perhaps smaller) return tomorrow than the (perhaps just as certain) larger return five years from now. In more technical language, decision-makers may well use a higher discount rate than that employed by traditional economic analysis (Brown, 1994, p 29). As a result, the benefits accruing from a regional scenario may well have to be 'front-loaded' in order to attract support.[7]

Secondly, it is important to recognize that various benefits and costs of regional actions would cut across three major levels: the individual, the national and the regional. Those arguing in favour of particular regional schemes concentrate upon the outcomes aggregated regionally. This is only natural, since justification for regional activities can only be made in this way. What must be accepted, however, is that the way in which the consequences are spread across these different levels could be important. Brown (1994, p 23) argues that 'member-state elites assess the utilities of regional cooperative schemes on personal, national, and regional bases, consecutively'.

Moreover, the calculations at the personal and national levels (indeed, at any level) may not wholly constitute monetized elements. Robson, for example, argues that 'money costs do not reflect true social costs because of such factors as unemployment, infant industry considerations, external economies and diseconomies and foreign exchange shortages' (1980, p 150). Indeed, this encourages the recognition that non-monetized elements may be registered and valued in different ways. There may, for example, be political kudos arising from leadership positions (the granting of an organizational home for a regional arrangement is but one example). Other activities, meanwhile, may have costs – for instance, the loss of national prestige associated with a transfer of decision-making power to a supranational body. The message is simply that the analysis must extend beyond those elements that can be monetized.

Though these warnings should not only be heeded by those studying the developing world (politics is politics the world round!), analysts focusing upon developing countries have nevertheless given it particular emphasis. Dash (1995, p 516), in his study of South Asia, maintains that:

> ...like many other Third World leaders, the South Asian leaders are also guided by a concern for their political and physical survival. Most regional and foreign policies promoted by the South Asian leaders are based on rational calculations; that is, the leaders pursue policies that are most likely to ensure their political survival, provide legitimacy to their rule, and keep them in power for a maximum length of time.

Østergaard (1993, p 36) quotes Vale (1982, p 33), who argues that 'the political practitioner has always to be conscious of the necessity to balance a commitment to the common (integrative) endeavour against the need to account to his constituency – local or national'. In a similar vein, Ravenhill (1979, p 229) is also quoted in the Østergaard article (1993, pp 36–37). He maintains that 'regional integration is frequently without enthusiastic domestic proponents: for politicians concerned with their national constituencies there are few rewards at the regional level, at least in the short term – the time horizons with which they must of necessity be concerned'. Indeed, Blomqvist (1993, p 53) seems to sum it up nicely by observing that:

> ...at the end of the day, individuals make the decisions, not nations. According to the so-called public choice school there is no such thing as 'national interest', but the term is used by politicians and bureaucrats as a rhetorical device for defending their personal interests (part of that interest may, of course, be what they perceive as a national interest).

An examination that is sympathetic to cognitive approaches, therefore, leads us to the widely recognized opinion that, to understand governments, the analyst must accept that there could well be particular demands (for example, electoral accountability) which may lead to assigning different, perhaps initially unexpected, priorities to different issues. For our study of regional cooperation in developing countries, therefore, we are led to believe that decision-makers in such countries could well place a higher than expected priority upon the certainty of short-term, tangible and political benefits arising from any regional cooperative effort.[8]

Country Interests

The preceding discussion about elite interests encourages us to redefine 'rationality' – that is, to focus upon developing countries' decision-makers, to increase the discount rates they employ and to place greater weight upon political factors in a revised calculation of interests. This would then lead to what we might consider a more political evaluation of the relative costs and benefits of regional proposals. Any temptation to place all faith in such an approach, however, should be resisted.

Instead, different costings will be employed in different instances. This is not only because of the unique cognitive frameworks used by individual leaders (for example, Langhammer and Hiemenz, 1990, p 116), but because different leaders will have unique pressures placed upon them. The message is simply that we should not assume that interests are defined by the state (that is, the government of the day), but could reflect the country more broadly. Though a narrower focus upon the former may well be more reasonable in developing countries (civil society is weaker in the South, but more about that below), it is nevertheless the case that divergent interests within the wider society could well affect the prospects for regional cooperation.

Accordingly, we widen the focus. Until now, we have been defining the elite as decision-makers, implicitly assumed to be political leaders, or perhaps close advisors. Even if we stay focused upon this idea of the 'elite', we see that other parties could be part of this elite, and therefore have some bearing upon the calculation of interests. For instance, there could be rivalries within government which generate different interpretations of interests. Particularly if the issue is not seen to be high politics by the nation's leader, then it may be left to the departments affected to formulate positions.[9] Their particular interests may well be affected by, in the famous words of Allison (1971), 'where they sit'. For instance, departments concerned with industry, planning, finance, energy and environment may have different perspectives on a proposal for regional cooperation. Depending upon how they are able to exercise influence, they will be able to affect outcomes to different extents.

Moreover, there may well be actors outside of government who are able to influence the understanding of interests. Axline presents a few ideas:

> Among various sub-national groups the ones most likely to take positions on regional integration are those which will be directly affected by the gains and losses from integration and which have some access to the political process at

the national or regional level. These include the local private business sector, both commercial and industrial, sometimes represented by chambers of commerce and manufacturers' associations; labour, both industrial and agricultural, possibly represented by individual trade unions and confedera-tions of unions and sometimes closely aligned with political parties; and the 'radical' left found outside the major political parties and often associated with intellectual circles. (Axline, 1979, p 52)

Blomqvist (1993, p 54) confirms their potential impact (with a particular emphasis upon the influence exerted by industrialists) by remarking that 'the "national interest" may in reality be the interest of a powerful lobby group'.

The conclusion is that when we think about interests, it could well be useful to go beyond the nation's leaders to explore the ways in which elements of government, and society more broadly, affect the understanding of interests.[10] It may well be that those who feel threatened by potential regional arrangements will be able to exert a disproportionate influence, exercising a veto and effectively blocking any change.[11] As Segal has argued, to be successful, an integration scheme 'must not threaten existing benefi-cial relationships or it must replace them with new ones' (Segal, 1967, p 263, quoted in Axline, 1979, p 61). As a result, even if all costs and benefits seem to yield a positive outcome for the country as a whole, those actors who come up short may effectively exercise veto power.[12] Our conclusion from this study of country interests, therefore, is that it is important to consider the ways in which different actors within society formulate their interests and influence decisions.

Equity

A discussion about the existence of particular benefits and costs – and how they are perceived by different players – leads on to a discussion about the distribution of these same benefits and costs across regional players. We therefore turn to a discussion of 'equity' in the process of implementing effective regional cooperative measures.

There is broad consensus in the literature that net gains need to be forthcoming in any potential set-up; all potential participants must also believe that the resultant gains (and any incurring costs) are distributed in a fair or just manner (for example, Aly, 1994, p 37; Axline, 1994b, p 189; Blomqvist, 1993, p 53; Johnson, 1991, p 4; Kisanga, 1991, pp 31 and 33; Langhammer and Hiemenz, 1990, p 67; and Robson, 1980, pp 151ff). Capturing the general sentiment of this literature, an OECD study argued that one of the reasons for the failure of many regional integration schemes was 'the inability to come to mutually-acceptable terms over the distribution of the costs and benefits from regional integration' (OECD, 1993, p 66).

Whatever scenario is developed, analysts argue that all participants must believe that the arrangements are taking their primary concerns into account and treating them fairly.

Theoretically, an unequal distribution of benefits (and perhaps a skewed imposition of costs as well) should be expected because of the so-called polarization effect of regional cooperative schemes. This is the hypothesis that the more advanced countries in the region will be better placed to reap the benefits of any cooperative arrangement. In the case of integration schemes, for example, capital 'will accumulate in more advanced countries, and labour will tend to migrate'. Therefore, 'synergy effects will work in favour of the relatively more advanced partner economies' (Langhammer and Hiemenz, 1990, p 15). Supporting this, though packaging the argument somewhat differently, Blumenfeld (1991, p 126) maintains that the gains from a regional cooperative scheme will be distributed 'in accordance with the two countries' relative bargaining powers' – suggesting that the strong are more likely to reap the benefits. Indeed, Axline argues that this will particularly be the case in developing countries:

> *Among underdeveloped countries disparities in the distribution of the gains from integration are likely to be greater, with some member countries being net losers. This is because of the effect of 'spread' and 'backwash' effects of integration, where the former, representing an outward spreading effect of the benefits of integration prevail in industrialized areas, and the latter, representing a clustering of the gains around the growth poles of the region tend to prevail in underdeveloped regions.*[13] (Axline, 1979, 14)

Regardless, the concept is not mere speculation since there certainly has been a skewed distribution of gains arising from past arrangements among developing countries. Gambari (1991, pp 9–10), for one, argues that the 'unhappy experience of the East African Community illustrates the dangers of a "free" common market that leads to the concentration of economic development in the member country (in that case, Kenya) which already has an established industrial base'. Hansen (1969, p 259), looking at both Latin America and Africa, maintains that the major issue was 'that of gains *relative* to those of one's partners, not as measured against any absolute standard'. Johnson (1991, p 8) reports similar opinions. When compensation schemes of some sort – development banks and the like to help the smaller partners – have been introduced (for example, Hansen, 1969, p 260; Langhammer and Hiemenz, 1990, p 60), an uneven distribution of benefits and costs has often still resulted.

This discussion has resonance with studies of international relations more generally. Many who study international society, particularly those who view it as anarchical, argue that the *absolute* benefits arising out of international cooperative efforts are not of paramount importance; instead, the *relative* benefits are crucial. This emphasis upon absolute versus relative benefits has parallels with the ways in which some analysts, and practition-ers, view international transactions as either *positive-sum* or *zero-sum*. To phrase it more colloquially, those emphasizing the importance of relative gains, and viewing international relations as zero-sum, look not only for what arrives in their lap as a result of regional cooperation, but also look over the shoulders of others and into their laps to see how they have profited from the new arrangements. They would rather that each person receives nothing new than have the smallest piece of a newly created pie.

The emphasis placed upon equity in many different ways encourages us to think about how the concept may affect the prospects for regional cooper-ation among developing countries. Indeed, it is really not enough to search for *the* equitable solution: just as we have seen that interests can be rationally defined in many ways, so too can terms such as 'equitable' and 'fairness'.[14] Some consideration, therefore, of how different actors would define a fair arrangement would seem to be an important part of any analysis of regional cooperation in the developing world.

Power

A discussion about a world consisting of ambitious players, each seeking to achieve maximum relative gains, leads to a more explicit consideration of power – the mainstay of much of the study of international relations (and therefore a potentially important factor for any study of regional coopera-tion). It is certainly inevitable that power – the ability for one actor to bring about a change in the behaviour of another – will be important in any inter-national activity. With regard to how its distribution affects the prospects for effective regional cooperation, a couple of propositions have been advanced.

On the one hand, some individuals suggest that asymmetries of power will be beneficial. This derives from the notion that the presence of a hegemon in the region – that is, an individual actor (usually a state) with a preponderance of resources – will be a positive thing. The hegemon will want the benefits from regional cooperation so badly that it will be willing to do whatever is necessary in order to secure them, even if that means bearing a disproportionate share of the costs of establishing the regional arrangement. It does so because it anticipates that it will accrue even larger benefits. These may be either tangible (perhaps economic prosperity or security) or intangible (perhaps prestige) (for example, see Kindleberger, 1986, p 8). Thus, the small countries in the proposed arrangement are allowed a free ride. This encourages Langhammer (1991, p 138) to conclude that 'large member countries tend to be exploited by small countries' in regional arrangements.

On the other hand, some suggest that the existence of a dominant player will do little to help the cause of cooperation at the regional level. From the perspective of the smaller actors in the region, they may try to hinder the efforts of the hegemon because they are fearful of establishing a dominant, neocolonial relationship of some sort. Or they may simply have seen depressing past experiences and not want to experience the backwash effects that are often the consequence of polarization in regional arrangements. Laszlo and colleagues (1981, p 18) are obviously concerned about this consequence of power asymmetries when they argue that they can 'give rise to the worry that the larger regional members will use their political power, or superior population size, to coerce the smaller into agreement (despite the possibility of establishing qualified voting procedures and other systems of checks and balances)'.[15] From the perspective of the powerful actor, meanwhile, the hegemon may not want to pursue regional cooperative efforts; it may feel resentment as it sees free-riders trying to reap the benefits without contributing to the costs.

Which effect will outweigh the other? It is, of course, difficult to know in advance (for example, Zormelo, 1995, pp 25–26). What appears to be beyond dispute, however, is that any analysis of efforts to forge regional arrangements, or the prospects for the same, needs to be aware of the ways in which power asymmetries may affect the dynamics of the process.

Homogeneity

The title of this section may seem somewhat misplaced. Are not some kind of differences among the countries of the world necessary in order to have regional cooperation? Indeed, why would they be interested in regional action if not to exploit parallels? This is certainly true; if all countries were identical, there would be little incentive for one country to interact with another, and thus little motivation for (or need for) regional cooperation.

We are, however, starting from the premise that the potential benefits of regional cooperation (often predicated upon some kind of difference within the region) already exist, at least in theory. As we consider the prospects for their successful realization, we take the next step: following the literature, we explore the hypothesis that a region must be similar in many ways in order to secure effective and sustainable regional cooperation.[16]

Of course, the extent of homogeneity within a region can be assessed by looking at a range of characteristics. The literature concerned with integration and customs unions, for example, highlights the need to have partner economies that are approximately the same size. Were they not, then 'difficult economic adjustments – especially labour dislocations – [could arise] that result in political strains in and between members' (OECD, 1993, p 31). Thus, such similarities might serve to eliminate the polarization effect.

We can move on and consider characteristics other than the size of the economy. We can look at other economic factors (for example, general level of development or policies regarding the same), more explicitly political characteristics (for example, congruency of governmental types, elite value parallels, even ideology in the broadest sense), social or cultural attributes (for example, systems of law or language), or technical traits (for example, harmonization of any particular set of standards) (see for example, Hettne and Inotai, 1994, p 3; Johnson, 1991, p 17; Langhammer and Hiemenz, 1990, p 13; and Schweickert, 1996, p 50). Therefore, we can take the message that comparing the attributes of the countries in the region – at these different levels – may be helpful in understanding the development (or not, as the case may be) of regional cooperative efforts.

Orientation

Orientation – the degree to which countries 'look at' each other – is another important element of regional cooperative efforts.[17] First of all, many argue that economic orientation has been an important precondition of successful integration schemes to date. Robson (1980, quoted in Kisinga, 1991, p 28), for example, emphasizes the extent to which the potential partners in an economic scheme are already conducting 'a significant proportion of their trade with one another'. Blomqvist (1993, p 66), furthermore, argues that: 'Economic integration does not seem to work in practice when the "natural" trading partners are found outside the integrated area.' Hettne and Inotai (1994, p 9) also place emphasis upon the orientation and compatibility of economic policies.[18]

Another interpretation focuses attention upon explicitly political orientation – in other words, looking at the extent to which political links among the countries already exist (Keohane and Nye, 1977; Axline, 1979, p 57). The rationale for this particular emphasis is that if a proposed arrangement is to make use of existing structures, then it is more likely to generate support. Alternatively, the requirement to create new structures further increases the extent of necessary (and difficult) change.

We can take this notion of orientation to a third level, draw upon more explicitly technical aspects, and look at the extent to which functional linkages already exist among the candidate countries. This is derived from the ideas developed under the functionalist and neofunctionalist studies of Europe and the European Communities.[19] This would lead us to investigate the extent to which, in a very general sense, the countries look at each other. How many cultural exchanges, for example, are already taking place (Blomqvist, 1993, p 56 hints at this)? Do the region's experts already communicate with each other (and thus perhaps already operate in what has been called an 'epistemic community' – see, for example, P M Haas, 1990)? Are people moving around the region, for either work or pleasure? Are, moreover, citizens calling themselves members of a particular region (remembering that regions can be socially constructed)?

Many have captured similar ideas within discussions about a regional civil society. Hettne and Inotai (1994, p 3), for example, argue that its presence is crucial:

> *The new regionalism...presupposes the growth of a regional civil society opting for regional solutions to local and national problems. The implication of this is that not only economic, but also social and cultural networks are developing more quickly than the formal political cooperation at the regional level.*

Indeed, a number argue that the people of the region themselves must want the arrangements, or else any arrangements that are formed will not be sustainable. Musoke's highlighting of the importance of 'Involvement of the people in the region; the people should see, feel and enjoy the fruits of cooperation' (1990, p 35) is representative (see also Aly, 1994, p 135; and Langhammer and Hiemenz, 1990, p 75). A recent study for the OECD (Mytelka, 1994, p 7) makes a similar case: Mytelka argues that there is a need 'to actively involve private-sector and other non-governmental economic actors in the design and launching of new forms of South–South cooperation, and for policies that foster networking for innovation among such actors across national boundaries in developing countries'.[20] Therefore, some mapping of the transactions already existing within the region would appear to be a good indicator of the prospects for successful regional cooperation.[21]

Linkage

The way in which the proposed arrangement on a particular issue may be linked to other issues could be important. Events and conditions seemingly unrelated to the issue under consideration may provide a window of opportunity, or may be in some way conducive to regional cooperation. Alternatively, they could also be unhelpful. Osherenko and Young (1989, pp 260–261), for example, argue that:

> *The state of the broader political environment is a key determinant of the prospects for regime formation in specific issue areas. Sometimes the political environment is conducive to efforts at institution building; it may even provide a powerful impetus toward regime formation, regardless of the content or coherence of specific proposals ... By the same token, the broader political environment may impose severe constraints on regime building in specific issue areas.*

Within the regional context, this may inspire more explicit consideration of how the issue is linked – economically, politically or even just psychologically – to other regional issues (and, of course, to developmental objectives more generally).

External Factors

This notion of linkage (or context) directs our attention to the way in which the landscape in the broadest sense – that is, the world – influences regional cooperation. More specifically, we should look at factors external to the region to see how other countries, actors and international structures might lend support to, or alternatively inhibit, regional initiatives.

In the analysis of regional cooperative efforts, many have at best under-valued, and at worst overlooked, such external factors (as noted by, for example, Nye, 1965, p 882, quoted in Hansen, 1969, p 70); and Axline, 1994b, p 190). Axline (1994a, pp 25–26) argues that the 'link between external actors and the process of regional cooperation is through their influence on the costs and benefits of regional cooperation as a whole and the distribution of these costs and benefits among member states, which will determine their perceived opportunity costs'. Indeed, given increasing globalization, it would appear that some focus upon extraregional factors would be crucial. Which, however, are most important?

Those who have considered external factors have usually paid most atten-tion to transnational corporations (TNCs) (Kisanga, 1991; and Vaitsos, 1978, pp 729–736); with considerable resources at their disposal, their impact could potentially be large. Vaitsos (1978, p 729) argues that they 'influence policies, participate in or even dominate policy implementation, and can become critical integrating or disintegrating forces in the pursuit of their corporate objectives'. There are different ideas about what might encourage, respectively, these integrating or disintegrating tendencies.

On the one hand, Østergaard (1993, p 37) proposes that 'TNCs may promote regional integration of the market type if the national markets are fairly small and if such firms were not involved, through parallel foreign direct investments, in these countries prior to cooperation' (see also, Kisanga, 1991, p 40; and Vaitsos, 1978, pp 733–734). On the other hand, if the TNC already has subsidiaries in two or more countries in the proposed regional arrangement, it could well have little interest in integration (Vaitsos, 1978, pp 732–733). Similarly, if cooperation will serve either to allow competitors new access to the region (Kisinga, 1991, p 40), to encourage indigenous entrepreneurs to establish fledgling operations, or simply to lessen the number of producers in the sector (Vaitsos, 1978, p 733), then the TNC may try to inhibit the process. Thus, the kind of cooperation proposed, along with the particular position of the TNC and its regional competitors (both present and potential), will be crucial for determining the attitudes of executives running the transnational business.

Other kinds of foreign investors – particularly those bringing in what is called 'portfolio investment' – may also be important. The growth of the so-called 'emerging markets' during the early 1990s is well documented (IFC, 1995) and there is little doubt that these emerging stock markets have been an important source of new finance for developing countries.

The purpose here is not to assess the relative merits of the stock market as an engine of economic development (for that, see Jefferis, 1995). Instead, the key point to recognize is that stock markets exist in the developing world, and that their relative buoyancy is important to a range of actors, both inside and outside of the particular markets. As such, the phrase 'what is good for the stock exchange, is good for the country as a whole' is accepted by many.[22] Therefore, the attitude of the market to regional cooperative schemes may well be important. If it is anticipated that the market will react favourably to a particular arrangement, then those who agree that the health of a country's stock market is important may work to facilitate implementation. The corollary, meanwhile, may also be true. As a result, our analysis should explore the possible reaction of 'the market' (which perhaps would be best made tangible by investigating the reactions of international fund managers) to proposed regional arrangements.[23]

Foreign governments are another set of actors which could impact regional cooperation (for example, Axline, 1994a, p 25; and Johnson, 1991, p 18). This is especially the case in the developing world, where the 'trade, aid and investment policies of [foreign] countries play a large role in defining the alternatives to regional integration and thus in determining the opportunity costs of participation for member governments' (Axline, 1979, p 56).[24] Again, the impact of advancing regional cooperation could be positive or negative. For instance, the government may back a cooperative scheme for the same reason as its proponents do – namely, that it will serve to advance development in the region. Alternatively, however, foreign governments may not want to see the region gain in either economic or political power, and thus may prefer a 'divide and conquer' strategy in its relations with the region's countries (Vaitsos, 1978, p 727).

A third area is international organizations (IOs). While some might consider IOs to be mere subsets or spinoffs of foreign governments (for that is, by definition, what an IO is comprised of), others accept that IOs may acquire their own identity (for example, E B Haas, 1990). As a result, there may be particular interests associated with the IO which may not be directly reducible to the sum of its parts (the interests of the respective member states). Accordingly, they merit some attention in any analysis of regional cooperation.[25]

Finally, international structures more generally warrant investigation. Some broader studies of international relations highlight the fact that outcomes are not only affected by agents but by structures too. Alternatively called regimes and institutions (and no doubt a host of other names as well), it is nevertheless the case that particular sets of norms and practices acquire some influence independent of any particular advocate.[26] Accepted rules of the game may be an expression that gives this more substance. However referred to, any study of regional cooperation would be wise to consider the extent to which external (perhaps global) institutions affect the dynamics of regional cooperation. This may include issues such as trade and debt (for example, Tsie, 1996, p 76), but also the elaboration of ideas about concepts such as development and globalization.

Capacity

Finally, any study of the prospects for policy implementation must give significant attention to capacity – that is, the ability of the relevant actors to carry out the actions necessary to realize the intended outcome. Indeed, discussions about the appropriate role of the state in the developing world, and its ability to carry out its core functions, are highly relevant here (see for example, Jackson, 1990). Given the significant challenges that already exist at the level of the state, when we add another level – the international – to the scenario, the potential for even greater capacity constraints increases. Such considerations encouraged de Melo and Panagariya (1993, p 15) to argue that 'when there was implementation, the terms proved to be too ambitious for the members' limited administrative capacities'. All cooperative arrangements will require at least some level of capacity among particular actors, often the state but others as well. At any rate, it is imperative that any analysis explicitly considers the organizational capacity of the participating entities.

APPLICATION TO THIS STUDY

The preceding review makes one thing clear: a wide range of factors has been identified in response to the question 'How might regional cooperation, particularly among developing countries, be encouraged and sustained?' Given this, the next task is to determine how this existing knowledge might be used in our own study when assessing the prospects for future climate change cooperation.

The discussion above has already presented some evaluation of the literature on regional cooperation among developing countries; namely, if a particular factor has been included above, then it is of at least some explanatory value. (The reader must recognize, of course, that the number of propositions related to the processes of international cooperation is limited only by one's imagination!) Beyond this, however, no assessment of the literature was undertaken.

As a result, we are left with propositions that may have attracted varying degrees of acceptance by scholars and practitioners. We also have propositions that may not be mutually exclusive – that is, two or more may interact with each other to offer the most useful explanation or indicator of future outcomes. Indeed, different propositions are concerned with different stages of the process of regional cooperation: some are concerned with the conditions that need to be in place before a regional cooperative scheme can even be contemplated, while others are recipes for the regional arrangements themselves. Consequently, the fact that the ten factors have been identified sequentially should not lead the reader to believe that they are necessarily of one kind.

They are, however, useful ideas about how we might tackle our investigation into regional cooperation among developing countries. Seeking explanation for particular phenomena, scholars use theory to guide their investigation. If they did not have theory, then they would not have any guidance with which to begin their study; they would have no idea where to look for answers. Consequently, their decision to begin at 'A' rather than 'B' would be determined by luck, or something similar. Though it would not necessarily matter where one started if one was going to be exhaustive in one's search, instances in which resources are unlimited are usually quite rare.

Theory, then, is used to guide the analyst in his or her study. In this study, theory – captured by the discussions in this chapter – will be used heuristically to structure the investigation in the rest of this book. The ten factors, summarized in Box 2.1, will be used to guide the search for the answer to: 'How can regional cooperation in southern Africa both promote the mitigation of global climate change and advance the region's development objectives?' Of course, it is certainly the case that the existing theory does not have a monopoly on truth; there may be particular factors that are unique to the case under investigation. That possibility is certainly accepted, and it is hoped that by not being excessively restricted by propositions – but, instead, simply guided by the same – such factors will find space in the studies we are about to undertake.

The other issue worth investigating here is the extent to which this chapter's discussion about regional cooperation (in its widest sense) relates to the book's ambition to understand climate change cooperation. It could be asserted that the latter is such a specific version of the former that only parts of our broader theoretical survey are relevant. Such an assertion, however, would be premature.

At this point, we do not know what shape the best form of climate change cooperation will take (however we might define best). It could be a

Box 2.1 Summary of factors potentially affecting the prospects for climate change cooperation

The ten factors which potentially affect climate change cooperation are:

- sovereignty;
- elite interests;
- country interests;
- equity;
- power;

- homogeneity;
- orientation;
- linkage;
- external factors;
- capacity.

variation of the classical economic integration – for example, the development of a regional market for energy-efficiency technologies. Alternatively, it could take a much more restricted approach – for example, project coordination in the form of the construction and operation of a regional hydroelectric facility.[27] Accordingly, it is important that we, at this point, do not unduly restrict climate change cooperation to particular forms of regional cooperation. Were we to do so, options that might serve to be most beneficial – on environmental, economic and social levels – could be prematurely jettisoned.

In conclusion, then, the subsequent chapters will try to make use of the discussion in this chapter so that the direction of our investigation is not determined simply by luck, but instead makes use of the travels that others have undertaken on similar journeys. At the same time, however, should interesting diversions appear, we will not resist the temptation to explore them.

ENDNOTES

1   International cooperation has been the subject of numerous studies. See, for example, Milner, 1992; and Young, 1989.
2   Following Keohane (1984, p 51), let us take regional cooperation to occur when actors in two or more neighbouring countries adjust their behaviour to the actual or anticipated preferences of others, through a process of policy coordination.
3   Park also identifies the Southern African Development Coordination Conference (SADCC) as a particular success. SADCC will be examined more closely in Chapter 3.
4   Somewhat related to this is the argument that political will must be present if regional cooperation is to be forthcoming. Though often identified as crucial (see, for example, Hazlewood, quoted in Blumenfeld, 1991, p 139; and Kisanga, 1991, p 3), it is usually not accompanied by a tangible explanation as to what this political will is (apart from the rather unhelpful conclusion that it was obviously present if the regional scheme ended up being successful, while definitely absent if the arrangement failed). Accordingly, it is mentioned here, recognizing its presence in the broader literature, but not explicitly pursued, since no guidance is provided by that same literature on how it might be useful, either analytically or prescriptively.
5   One reason might be that a stronger power is forcing others to consider cooperative action. The particular plan may not yield benefits for the threatened country, but is a more preferable alternative (the lesser of all evils) than that of non-cooperation. Recognition that regional cooperation should always be considered in the context of other plausible options is obviously crucial. We do consider the importance of power in this chapter.
6   Supporting this, Segal identifies one of the conditions for a successful integration scheme: 'it must not threaten the bases of support of existing national political elites' (Segal, 1967, p 263, quoted in Axline, 1979, p 61). Jinadu (1990, p 11), however, warns us that it may be difficult to discover the attitudes of the elite towards particular integration schemes.
7   Related to this, the more tangible the set of benefits, the greater the likelihood that the associated proposal will be supported. Similarly, the simpler the proposed

scheme, the greater the likelihood that it will attract support. Why? Simply because there is something comforting about proposals that are either conceivable, familiar or intuitively attractive.

8   This particular factor is not distinctively regional – that is, the successful implementation of national mitigation proposals may well be affected by consideration of elite interests as well. Indeed, many of the ideas we raise in this chapter might be equally applicable at the national level.

9   A number of individuals argue, however, that virtually every issue contemplated by developing country governments will be considered to be high politics. Almost 30 years ago, Hansen (1969, p 260) argued that the evidence clearly suggested 'that, for less developed countries, economic integration is better conceptualized in terms of high politics than its welfare equivalent in Western Europe'. Much more recently, Axline (1994b, p 181) echoed this sentiment.

10 This is one of the key messages emerging from Axline's study (1994a, p 28).

11 Recognize the powers of inertia: it is far easier to continue on the path that one is on than to generate support for a programme of change.

12 Others argue that even if the aggregated outcome is a net benefit for a single actor, that actor may still not choose to proceed. The reason is because many people are risk averse. They give potential losses disproportionate importance. For example, a simple summation of consequences would suggest that the outcome of hypothetical regional action 'A', which results in a gain of 100 widgets and a loss of 50 widgets, would be preferable to regional action 'B', which results in a gain of 30 widgets. This, however, may not be the case. 'Laboratory research reveals that people are generally more concerned about loss than gain. Pain is a more powerful motivator than pleasure. In framing choices, people weight potential losses more heavily than gains' (Brown, 1994, p 24). Leaders, therefore, may prefer 'B' to 'A'.

13 Hansen (1969, p 256) had earlier argued similarly, relying upon Balassa (1961, p 204).

14 See, for example, the discussion in Rowlands (1997), where I consider how different understandings of international fairness and justice generate vastly disparate policy prescriptions for industrialized countries on climate change.

15 There is not necessarily a guarantee that the most powerful state in the region will pursue arrangements that will be beneficial to all. As such the hegemon may be acting malevolently, rather than benevolently (as suggested above). For more on this, see Snidal, 1985.

16 To reconcile this apparent paradox – that is, the simultaneous presence of both similarities and differences – we note that the differences often appear to focus upon the countries' endowments, while the similarities are more concerned with the countries' consciously constructed social institutions. Consequently, there is nothing to suggest, at least on a theoretical level, that such similarities and differences cannot coexist.

17 This is akin to Deutsch's ideas about responsiveness. See Deutsch, 1957, quoted in Zormelo, 1995, p 20.

18 There are clear hints of a tautology here; it seems that it is being suggested that 'in order to have regional cooperation, we must have regional cooperation'! This, of course, would be ridiculous. Instead, the message that the literature appears to be sending is that 'you must walk before you can run and, indeed, crawl before you can walk'. Because many ambitions concerning regional cooperation appear to be akin to running, the analyst should see – to keep with the metaphor – if any crawling (at least) already exists!

19 See, for example, Haas (1958). There is a danger in using explicitly 'Northern theories' to develop ideas about 'Southern phenomena'. Given that the paradigm case of regional cooperation – that is, Europe – comes from the North, it is obviously an issue that must be confronted in this study. The warnings that have been issued about the danger of transferring such ideas are well heeded. Nevertheless, since there are some similarities – particularly that all countries exist within a common international system, based upon the twin concepts of sovereignty and territoriality – we would be ill-advised to ignore them completely.

20 Conceivably, orientation may also be a hindrance: the animosity from existing or past interactions could serve to hamper current efforts. We will be attentive to such a possibility.

21 There are those who look to the already existing transnational civil society links (for example, the functionalists noted above), while others hint that there need be only existing civil societies within countries. Hoffman, for example, regarded 'fully integrated units ("political communities") and pluralistic social structures within them as essential requisites for successful regional integration' (quoted in Hansen, 1969, p 262).

22 The stock market can serve as an important source of international capital. Notwithstanding the dangers of volatility associated with such inflows, a stock market can still play an important role in strengthening the balance of payments and thereby contributing to the expansion of international trade and supporting economic development more generally. In many cases, countries' leaders would like international investors to view their markets favourably.

23 The attitude of domestic fund managers would also contribute to the thinking of the market as a whole and will thus not be ignored. We should not, however, strictly call this an external factor. What we call it is not particularly important – what is crucial is that it receives the attention it deserves.

24 Though their comments are somewhat dated, the general essence of what Laszlo et al (1981, p 18) argue may still hold true:
    ...dependency relationships between developing countries and one or more industrial nations which, whether the legacy of colonialism or the result of more recent developments, produce a reluctance in developing countries to evolve alternative regional relationships for fear of jeopardizing their unwelcome but seemingly necessary North–South ties (notwith-standing the fact that regional integration could replace them with a wider range and more equitable set of international economic relationships).

25 Axline (1994a, p 9) is one of the few to highlight their potential role.

26 For a general discussion about the influence of institutions, see North (1990). For a discussion applied to international issues, see, for example, Cortell and Davis (1996).

27 As such, we also recognize that ideas about what promotes regional cooperation (that is, the ten different factors identified above) may have different salience for different kinds of regional cooperation.

Chapter 3 | Regional Cooperation in Southern Africa

Ian H Rowlands

INTRODUCTION

The previous chapter presented ideas about what might drive, or alternatively inhibit, regional cooperation among developing countries. The purpose of this chapter is to review efforts to promote regional cooperation in southern Africa, as well as to survey its potential. Thus, while the previous chapter provided the conceptual foundation for our study, this one gives the empirical base.

The chapter is divided into two main parts. The first examines two of southern Africa's principal regional organizations: the Southern African Development Community (SADC) and the Common Market for Eastern and Southern Africa (COMESA). For each, their major activities are reviewed, and efforts are made to explain their relative successes and failures. The second part of the chapter presents a survey of contemporary southern Africa: guided by the framework laid out in Chapter 2 of this book, we discuss whether the conditions for regional cooperation, at a fairly general level, are in place. Together, the two parts of this chapter are designed to give the reader a better understanding of the experience with, and potential for, regional cooperation in southern Africa. Our attention then begins to turn more specifically to climate change cooperation at the end of this chapter. In a brief appendix, the history of regional cooperation in power sharing is reviewed, and a summary of the present state of affairs is presented.

REGIONAL ORGANIZATIONS IN SOUTHERN AFRICA

International cooperation in southern Africa is as old as the region's nation states themselves. Moreover, even during colonial times, there were substantial relations among areas that now belong to separate countries (Blumenfeld, 1991). However, the full history of regional cooperation in southern Africa – indeed, even the full history of regional *organizations* in southern Africa – is beyond the scope of this chapter (see, for example,

Söderbaum, 1996). Instead, we initially focus upon the two regional organizations with the widest membership.

The Southern African Development Community (SADC)

Primarily at the initiative of Botswana's President Sir Seretse Khama, the first meeting of the Southern African Development Coordination Conference (SADCC) was held at Arusha, Tanzania, in July 1979.[1] Building upon the cooperation that was developed during the war of liberation in Zimbabwe, representatives from the five frontline states – Angola, Botswana, Mozambique, Tanzania and Zambia – attended, as well as officials from various donor countries and international agencies. One year later, at its first summit meeting in April 1980 at Lusaka, Zambia, SADCC was enlarged to include representatives of Malawi, Lesotho, Swaziland and Zimbabwe. These nine countries would constitute the membership of SADCC throughout the 1980s; 1990 would see the entry of newly independent Namibia as the tenth member.

What originally brought these people together – in addition to a shared desire to promote development in the region – was a common aspiration to lessen their respective countries' dependence upon the apartheid regime in South Africa. More specifically, four main goals were identified (quoted in Meyns, 1984, p 203):

- the reduction of economic dependence, particularly, but not only, on the Republic of South Africa;
- the forging of links to create a genuine and equitable regional integration;
- the mobilization of resources to promote the implementation of national, interstate and regional policies;
- concentrated action to secure international cooperation within the framework of our strategy for economic liberation.

In this way, SADCC was encouraged to be different from the standard case of regional cooperation – not least because of its more modest aspirations. Moreover, instead of aiming for economic integration, the organization would try to generate regional cooperation on specific industrial projects. To do this, it would engage in 'coordination, harmonization, and selective intervention' (Aly, 1994, p 83). Anglin (1983, p 692) expands upon SADCC's break from the norm by identifying four areas in which it was innovative:

> ...first, the respect paid to the sensitivities of members to infringements on their national sovereignty; second, the degree to which responsibility for opera-

tional programs has been devolved to national governments; third, the balance of power between the political and bureaucratic organs; and fourth, the institutionalization of external dependency.

This 'pragmatic approach' (Leys and Tostensen, 1982, p 67) was deemed to be most appropriate, particularly given the then recent experience of regional disintegration in the shape of the East African Community.

In January 1992 – in light of the changes occurring not only within South Africa, but the broader international environment as well – a meeting of the SADCC Council of Ministers agreed that the organization should be transformed from a Development Coordination Conference into a fully fledged Development Community. Thus, the Southern African Development Community (SADC) was born in August 1992 with the signing of a treaty in Windhoek, Namibia (SADC Treaty, 1993). With its entry into force on 5 October 1993, the ten countries committed themselves to much fuller cooperation: 'deeper economic ... integration', 'common economic, political and social values and systems', and 'strengthened regional solidarity, peace and security' (SADC Treaty, 1993). Although many of the structures that were part of SADCC continued to exist in SADC, one important innovation was this commitment to economic integration. With the entry of South Africa (in 1994), Mauritius (in 1995), the Democratic Republic of the Congo (in 1997) and the Seychelles (in 1997), SADC's present complement of 14 was reached.

SADC continues to be active on a range of issues: during its 1995, 1996 and 1997 summits, protocols on shared waterways; combating drug trafficking; energy; transportation; communications; education; and mining were signed; so too were declarations on gender equality and land mines. Additionally, SADC has been conducting greater activity in the realm of trade and economic integration. Building upon the principle articulated in the organization's 1992 founding document, SADC leaders, at the August 1996 summit, agreed to create a free trade area within eight or nine years. SADC's current main areas of interest – along with the respective countries' coordinating responsibilities – are detailed in Box 3.1.

Evaluating SADC(C)

According to many, SADC(C) has been one of the most successful regional organizations ever to operate in sub-Saharan Africa, perhaps even the entire developing world (see, for example, Green, 1990, p 108). As evidence of the high esteem in which it is held, Foroutan (1993, p 250) reports that: 'SADCC has been considered by many inside and outside the [African] region as a successful example of [regional integration] to emulate elsewhere in [sub-Saharan Africa]'. The Organization of African Unity, meanwhile, has hailed SADCC as 'the most successful regional grouping in Africa' (quoted in Maasdorp, 1992, p 13). Usually accompanying such general

Box 3.1 Sectoral responsibilities for SADC members (as of August 1996)

Country	Sectoral Responsibility
Angola	Energy
Botswana	Livestock production; animal disease control; agricultural research
Lesotho	Environment and land management; water
Malawi	Inland fisheries, forestry and wildlife
Mozambique	Transport and communications; culture and information
Namibia	Marine fisheries and resources
South Africa	Finance and investment
Swaziland	Human resources development
Tanzania	Industry and trade
Zambia	Mining; labour and employment
Zimbabwe	Food, agriculture and natural resources

Source: SADC-USA, 1998

applause are plaudits for two particular elements of the body's work: its role in developing the region's transportation infrastructure, and its ability to secure resource commitments from international donors.

With respect to the former, the transport and communications portfolio has certainly been an active one, by far the largest of all of the SADC(C) areas of interest. A 1988 report revealed that transportation and communication accounted for well over one half of the total value of SADCC's planned projects, and over two-thirds of the money that had actually been secured (reported in Gambari, 1991, p 92). The Beira Corridor development – 'a 300-kilometre strip running from Beira on the coast of Mozambique to the border of Zimbabwe and containing Beira port, a railway, a road, an oil pipeline, an electricity line and a number of development projects' – is perhaps the most successful and well known of all of the projects within the portfolio (Foroutan, 1993, p 250). The activities also produced measurable results: '80 per cent of transit traffic from the six landlocked SADCC countries which had passed through South African ports in 1980 reversed to become 60 per cent going through non-South African SADC ports ten years later' (Kibble et al, 1995, p 47).

Related to this, the ability to secure resources from external actors has been another strength of the organization. Cooperating partners (as international donors are referred to in the language of SADC(C)) have substantiated their rhetorical interest in the body by financing over 90 per cent of all of the SADC(C) projects (Foroutan, 1993, p 250), including, in absolute terms, US$4.58 billion in the transport and communication portfolio alone (Mwase, 1994, p 35; see, also, Langhammer and Hiemenz, 1990, p 47). Donors, for their part, have seen SADC(C) as an effective means of promoting development in the region, particularly in the face of the apartheid regime in South Africa (for example, Kibble et al, 1995, p 47).

Though the most noted, observers do not restrict their praise to these two aspects of the organization's work. Maasdorp (1992, p 14) argues that SADCC has also 'inculcated among its members a habit of co-operating and ... it has forged a Southern African identity'; Tjønneland (1992, p 109) also talks of the emergence of a 'regional mentality' that SADCC has fostered, while Kibble and colleagues (1995, p 47) put 'helping to establish a regional identity' at the top of their list of achievements. The continuing interest in its annual meetings – on the part of both leaders and media – is further testament.

There are also, however, those who argue that SADC(C) has not even been able to meet its own relatively modest goals. With respect to development – always a prominent aspiration – most of the region's states experienced economic decline after 1980. These trends, critics maintain, could not be reversed by simple project coordination (Gambari, 1991, p 93).[2]

With respect to the goal of lessening dependence upon South Africa, meanwhile, some measures suggest that the SADCC experience was not particularly positive. A SADCC report from 1985, for example, noted that since 1980 the region had become still more dependent on South Africa for its trade outlets, and the 1986 summit meeting, 'although it recommended the adoption of economic sanctions against South Africa, failed to establish a timetable for doing so' (*Africa South of the Sahara*, 1995, p 117). Langhammer and Hiemenz's (1990, p 47) examination of the late 1980s suggests that this trend continued:

> *The record of summit meetings of SADCC during the period 1985–1988 suggests that the political objective has not yet been fully shared by all members to the same extent. Countries maintaining strong economic relations with South Africa in trade, capital transactions and labour migration like Malawi and Lesotho abstained from summit meetings. Furthermore, even without these two members, sanctions against South Africa as the main political point on the summit agenda could not be agreed upon unanimously.*

Gibb (1994, pp 224–225) also shows how SADCC countries' dependence upon South Africa increased after 1985, at least when using trade as a measure. He claims that there 'is little doubt that the ten SADCC states, both individually and collectively, were more dependent on South Africa in 1992 than they were in 1980' (Gibb, 1994, p 225).

Criticism has also been levelled at the poor commitment that the members themselves have shown to SADC(C) (for example, Tjønneland, 1992, p 109). While the fact, noted above, that the organization has been able to secure large amounts of external funding can certainly be viewed as beneficial, nevertheless, given that only one tenth (approximately) of support for SADC(C) projects came from local sources, any delinking from South African dependence may have simply led to increased donor dependence. Tjønneland (1992, p 109) also argues that the:

Problems and challenges have intensified since the latter half of the 1980s as SADCC has attempted to move beyond the 'easy' forms of coordination in transport and other infrastructure to embark on a new phase of cooperation: production, trade and related sectors and the wish to take on issues of coordination, even harmonization, of macro-economic policies.

Even before this time, internal cooperation among SADCC members may have been 'very fragile' (Langhammer and Hiemenz, 1990, p 47).

Perhaps one of the most stinging rebukes comes from Hawkins (1992, p 105), who believes that 'most of the regional infrastructure attributed to SADCC pre-dates it and would have been developed and rehabilitated even without SADCC'. This leads him to conclude that 'if SADCC were to be disbanded tomorrow, it would leave behind little of any permanence other than a large new secretariat building in Gaborone' (1992, p 105). Additional reflection upon SADC's more recent achievements and failures will be undertaken later in this chapter.

Explaining SADC(C)'s Performance

Given the mixed assessment of SADC(C)'s experience, it appears that there is a need to explain both particular successes along with particular failures. With regard to the former, it is widely agreed that the modest and pragmatic approach that the organization has adopted has contributed to the benefits that it has brought to southern Africa. Curry (1991, p 22), for example, argues that 'SADCC has been reasonably successful because it has tried to do relatively little'. In particular, he highlights the fact that it has avoided the trap of trade integration (Curry, 1991, pp 17 and 22) – a black hole that has swallowed many cooperative efforts in the developing world.

Success can also be explained by the fact that SADC(C) has always recognized and accepted the importance of its individual members' sovereignty (Gambari, 1991, p 87). It has not tried to develop a large central machinery which would develop an independent identity and capture resources for its own use (see, for example, Curry, 1991, p 22). Moreover, the generally common perception of 'South Africa as enemy' is offered as another reason for SADC(C)'s achievements; it was accepted that unity was crucial (for example, Gambari, 1991, p 87). Finally, the interest and support of the international collaborating partners has also been crucial.

These arguments echo different parts of the more general conceptual arguments developed in Chapter 2 of this book. Certainly, the concern for sovereignty is clear: SADC(C)'s experiences lend support to the general proposition that regional cooperative schemes have greater chance for success if they do not infringe upon the ability of members to exercise sovereignty. Ideas about elite values also resonate with this history: among SADCC members, the South African regime was equally abhorred by most. So too does the impact of external forces – in this case, foreign governments.

What is also required, however, is an analysis of the apparent weaknesses of SADC(C). Aly (1994, p 86) identifies a number of reasons for the organization's relative failure, dividing them into internal and external. Among the former, he identifies macroeconomic policy-making (on the part of individual countries) which did not sufficiently take into account regional coordination; external reasons include destabilization by South Africa, 'which has cost [US]$60–$90 billion since 1980'. Aly (1994, pp 86–87) also claims that the 'remote causes' of SADCC's poor performance can be traced to two major factors: 'a small and weak secretariat and heavy reliance on external assistance'.

For their part, Langhammer and Hiemenz (1990, pp 47 and 48) maintain that, though SADCC members were bound by a common aversion to apartheid, they nevertheless held differing ideologies. These, they note, 'should not be underrated'. Concern among members about the potentially negative consequences for national assistance receipts, should regional initiatives be supported, is also hinted at by some. Finally, a whole range of external factors have been identified: 'severe drought in some years and floods in others, war and civil unrest in Angola and Mozambique with severe spillover effects on Malawi, Zimbabwe and Zambia, depressed primary produce prices (until 1987), sluggish world economic growth during the first half of the decade' (Hawkins, 1992, p 108).

Again, after reflecting upon our conceptual investigation in Chapter 2, we discover that these explanations parallel some of the general theoretical propositions advanced there. More specifically, if the above explanations are accepted, then the relative importance of elite interests – particularly the need to compare regional benefits with national gains – is highlighted. Moreover, the significance of homogeneity, across a variety of levels, is also given support. And finally, the SADC(C) experiences suggest that external factors can influence relative success or failure. Emphasised by commentators in this instance is the importance of foreign governments (in particular, South Africa) along with broader international structures (in particular, the condition of the world economy).

The Common Market for Eastern and Southern Africa (COMESA)

With its roots in 1960s' initiatives launched by the UN Economic Commission for Africa (ECA), representatives from 12 states signed, in March 1978 at Lusaka, Zambia, a Declaration of Intent and Commitment on the Establishment of a Preferential Trade Area for Eastern and Southern African States. This was the first step towards creating a new body, which was finally realized at a summit conference at Lusaka in December 1981. Representatives from nine of the 18 states attending signed a treaty that established the Preferential Trade Area for Eastern and Southern African

States (PTA). At the same location six months later, PTA was formally inaugurated. With membership open to all countries within the ECA's eastern and southern African subregion, different states joined the PTA at varying times and with varying levels of commitment (Anglin, 1983, pp 688–691).

The purpose of PTA was: 'to promote commercial and economic co-operation in the region in order to transform the structure of production; to promote intra-trade; to develop industry; to co-operate in agriculture; and to achieve a common market by 1992' (quoted in OECD, 1993, pp 50–51). The means to achieve this were, initially, tariff reductions for a restricted range of commodities. These – which appear on what is known as the Common List – originally consisted of 200 items, but by 1993 numbered 319 (Shepherd, 1993, p 61). This movement towards free trade was to be followed by a regional customs union (Aly, 1994, p 31).[3] In reality, however, members were unable to keep up with the schedule outlined in the original PTA treaty, and the deadline for the first phase was pushed back to the year 2000. Even by the end of the 1980s, however, progress towards this goal appeared to be well behind schedule (Aly, 1994, p 31).

Just as SADCC remade itself SADC during the early 1990s, so too did PTA experience a metamorphosis: in November 1993, the member states signed a treaty transforming the PTA into the Common Market for Eastern and Southern Africa (COMESA). COMESA has the following aims: 'a full free trade area by 2000; a customs union with a common external tariff 10 years after COMESA comes into force; free movement of capital and finance; a payments union and freedom of movement of people' (quoted in *Africa South of Sahara*, 1995, p 126; see also COMESA Treaty, 1994). The agreement came into force – and so the organization was officially born – on 8 December 1994, after the necessary 12 states had ratified the treaty. At present, about 20 countries claim membership in COMESA (including those that have not ratified the treaty). Notable exceptions, however, include two SADC member states: Botswana and South Africa.

Of late, COMESA has had a fairly difficult time; indeed, it has experienced problems since, quite literally, day one. At its official launch in Lilongwe, Malawi, in December 1994, only 13 of the potential 23 members attended. There has been little improvement since:

> For the past two years Comesa has not been able to secure a venue for its annual summit. Namibia had initially offered to host the 1995 summit but it withdrew the offer in January citing unspecified 'security fears' for the heads of state who would attend the gathering. Sources say that [COMESA Secretary General Bingu wa] Mutharika approached Lesotho as an alternative venue, but the small kingdom declined to host the event citing financial commitments arising from the hosting of this year's SADC summit. Mutharika had promised to convene a summit before May but no one amongst its 23 member states would host it. (Gumende, 1996c, p 27)

Moreover, Gumende goes on to report that non-payment of fees is threatening to 'grind the organisation's programmes to a halt' (Gumende, 1996c, p 27).

The end of 1996 and the beginning of 1997 brought additional difficulties. In November 1996, a Mozambican representative announced the country's intention to resign from the organization, while a month later, an official from Lesotho announced his country's withdrawal: each cited the overlap between COMESA and SADC as one major motivation (Dludlu, 1997). Early 1997, meanwhile, saw the suspension of COMESA's secretary-general, Bingu Wa Mutharika. He was to be investigated for alleged misuse of funds at the organization's secretariat in Lusaka (Kayaya, 1997). Indeed, tension between COMESA and SADC – as each tried to shape or dominate the regional agenda – has been evident during the past few years (for example, Gibb, 1997, p 84).

Evaluating PTA/COMESA

Commentators identify some relative successes for PTA/COMESA. An OECD (1993, p 33) study, for example, identified the following as achievements: '[the] Reserve Bank of Zimbabwe acts as the multilateral clearing house; some tariff reductions [have been] introduced for a limited range of commodities; PTA travellers' cheques [were] introduced in 1988 to facilitate monetary transactions'. Maasdorp (1992, p 17) also highlights the increased use of PTA travellers' cheques and the reduction of some tariffs; he further notes that 'achievements thus far have been ... a simplification of customs procedures, easier cross-border transport, and the staging of an annual trade fair as well as meetings of buyers and sellers'. Tjønneland (1992, pp 105–106) and Mwase (1994, p 31) also cite some or all of the above. The secretariat itself, meanwhile, argues that 'intra-PTA trade grew at an annual average of over 8% after 1985, reaching a total volume of [US]$1.7 [billion] in 1992' (quoted in EIU, 1996a, p 64).

On balance, however, assessments have been much more critical than these comments might suggest (for example, Calland and Weld, 1994, p 13). As already hinted at, PTA/COMESA has been more successful in articulating plans than in actually implementing them: not only is this supported by reference to general delays about tariff reductions, but specific instances also exist. Langhammer and Hiemenz (1990, p 49), for example, report that '[g]overnmental decisions to accept a single Road Customs Transit Document in the PTA and to introduce a so-called PTA Third Party Motor Vehicle Liability Insurance Scheme for facilitating transit trade are simply not passed down the ranks so that arbitrary decisions ultimately rest with the customs posts'. In many cases, the PTA has offered few instruments to facilitate the rather specific goals (OECD, 1993, p 51). Moreover, even in those instances where tariffs have been reduced, 'non-tariff barriers have not yet been approached and remain decisive in relation to trade' (Robson, 1990, p 136; see, also, Tjønneland, 1992, p 106).

Quantitative data offer additional indictments. An OECD study (1993, p 46) found that between 1980 and 1988, the degree of openness (a measure of the ratio of trade to GNP) fell from 40.2 to 38.2. Between 1980 and 1990, meanwhile, the percentage share of intraregional trade (as a share of total trade) fell from 7.6 per cent to 5.9 per cent for member countries (Foroutan, 1993, p 247). Development, more generally, also stagnated during the 1980s, with per capita GNP rising by only 0.07 per cent (OECD, 1993, p 38). Finally, Langhammer and Hiemenz (1990, p 49) suggest that interest in the organization has simply not been present: 'no more than 50 per cent of Heads of States participated in each of the annual summit meetings between 1985 and 1988'.

Explaining PTA/COMESA's Performance

Many argue that PTA/COMESA was destined to perform poorly for a variety of reasons. For one, the incredible heterogeneity within the region doomed it to failure (see for example, Tjønneland, 1992, p 107). Consider, for instance, the size of the economy as but one comparative characteristic. The Comoros had, in 1981, an income of only US$1.3 million, while the figure for Kenya was US$7400 million (OECD, 1993, p 51). An OECD report argued that this 'wide dispersion in the levels of economic development suggested a priori that political problems could arise in the distribution of the gains from trade (and adjustment) even if they were small' (OECD, 1993, p 51). Geographically, meanwhile, the organization consists of both landlocked and island states; and though all members in the region were identified as eastern and southern Africa, many were separated by vast distances (and even by non-members).

In addition to this heterogeneity, a degree of homogeneity has also been advanced as a reason why the organization has had relatively little success so far. Within the grouping, there are only two relatively diversified economies, Kenya and Zimbabwe, with the rest being 'highly dependent on one or two commodities for their export revenues. The ensuing lack of complementarity...[has] thus far stalled all efforts towards any meaningful and effective integration' (Foroutan, 1993, p 249). The absence of mechanisms to redress the redistributive problem has also generated friction among members (Tjønneland, 1992, 107).[4]

Common commitment to the organization's objectives has sometimes been lacking: '[It] is not at all clear that the members of the group shared a common perception of the role of regional integration in the development process nor is it clear that they shared a commitment to such a strategy' (OECD, 1993, p 51). The fact that many countries not only belonged to PTA/COMESA, but to other regional bodies – with overlapping memberships – meant that the organization did not always receive the highest priority (Carrim, 1994, pp 8–9).

As was the case in our exploration of SADC(C)'s reasons for success and failure, we turn back to Chapter 2, in order to see if any of the propositions advanced there have been echoed in explanations for PTA/COMESA's performance. One that immediately appears relevant is homogeneity: the diversity of countries involved in the organization does seem to have prevented it from achieving as much as it otherwise might have done. Furthermore, our analysis suggests that the absence of comparable elite values (or even country interests) – perhaps not surprising given this heterogeneity – may have been another key explanation for relative failure. Finally, the issue of equity appears to have played a role as well.

Though we conclude our review of the region's organizations at this point, we will return to the issue in the final section of this chapter; we will then consider how the explanations for the performance of both SADC(C) and PTA/COMESA should be taken forward in the rest of this book.

THE PROSPECTS FOR REGIONAL COOPERATION IN SOUTHERN AFRICA

We now turn our attention explicitly to the contemporary situation in southern Africa. In this second section, guided by the framework laid out in Chapter 2, we will investigate most of the propositions presented there. If the framework that we have developed has some value, then the subsequent analysis should provide us with some ideas about the prospects for any kind of regional cooperation in southern Africa. This may prove useful when we turn to the explicit case of climate change cooperation in the rest of this book.

Sovereignty and Statism

Building upon the first three conditions identified in Chapter 2 – that is, the importance of sovereignty and the significance of both elite-level and wider country interests – we first consider the potential significance of such factors in the southern African context. A review of the literature suggests that they may well be crucial.

Many analysts have emphasized the importance of the state in Africa during the post-colonial period (for example, Langhammer and Hiemenz, 1990, pp 60–61). Directing attention more specifically to southern Africa, Thompson (1992, pp 139–140) argues that:

> *In the postindependence period of the 1970s and 1980s, the Southern African region was caught up in etatism, looking to the state to solve all problems. This triumphal enthusiasm was perhaps understandable in that many of the states were not under majority rule until after long and bloody wars. The state was worth fighting for, as the newly independent govern-*

*ments provided health care, education, literacy and water supplies to the
population for the first time. Further, the leaders and people both knew that
the economies were not yet in their hands; independence did not erase economic
linkages established over 100 years of colonialism. Therefore, the state was
exalted to unprecedented heights.*

Vale (1996, p 365) also supports this, more recently arguing that: 'For all
their anticipation of a regional future characterized by harmony, the states
of the region remain caught by the discourse of nationalism and the security
concerns that this predetermines.'

Though this should not suggest that the influence of the state will be all
determining – particularly given the downsizing of the state that is occurring
in many parts of the region – it reminds us that the ways in which proposals
will affect the state could still be important. Indeed, this has already been
shown during recent trade negotiations in southern Africa, when 'the clashes
between national and regional interests have become more pronounced'
(SARDC, 1997). Though many southern Africans are increasingly viewing
activities through some kind of regional lens, state and country interests
could still be important to the prospects of any regional mitigation option.[5]

Equity

Odén (1993, p 18) argues that equity issues, and the need to balance costs
and benefits, are at the core of the southern African 'regional project'. One
need only look to the founding document of SADCC to see the importance
of equity: the desire to create 'genuine and equitable regional integration' is
the new community's second enumerated goal (quoted in Meyns, 1984, p
203). The 1992 SADC Treaty, moreover, not only recorded the members'
determination for 'harmonious, balanced and equitable development of the
Region' (Preamble), it also identified 'equity, balance and mutual benefit' as
one of the community's five principles (SADC Treaty, 1993).

Cynics might well suggest that concern for equity is simply an acceptable way for states to fight for their own national interests; as such, it adds
nothing to the discussion of sovereignty and statism undertaken above.
Granted, it is questionable what motivates the calls for equity. What,
however, is beyond challenge is that a desire for equity has consistently been
integral to any recent action in southern Africa. As a result, the ways in
which the benefits and costs of regional mitigation plans are distributed
around the region merit some consideration. We will therefore give them
some thought in subsequent chapters.

Power

Power – the means whereby one actor can exert influence over another – can be exercised in many different ways. Many different measures of power can therefore be envisaged. However, no matter what measure is selected for investigation in southern Africa, the chances are extremely high that it would reveal South African dominance within the region. For example, in terms of economic activity, the country accounts for over three-quarters of all goods and services produced in southern Africa (UNDP, 1997). Given such a concentration of power, we could speculate that South African attitudes towards regional cooperation – and others' attitudes towards South African participation in the same – will largely influence the prospects for successful implementation of any regional arrangement.

Going beyond mere speculation, experience reveals that anxiety about South African dominance has characterized regional activity throughout this century. SADCC was, of course, originally formed because of grave concern about the impact that South Africa was having upon its neighbours to the north. Moreover, the emergence of a democratic government at the tip of the continent has not meant that all such concerns have completely disappeared. One commentator at SADC's 1996 consultative conference, for instance, reported that:

> *South Africa was publicly accused of repeatedly making verbal commitments to regional solidarity, but in practice continuing to look after its own interests, exploiting its neighbours' trade liberalization policies while keeping its own economy closed and sidelining regional organisations. Some conference delegates said that South Africa had been arrogant and bullying in its attitude to its regional partners.* (EIU, 1996b, p 19; see also Gumende, 1996b)

Concerns about regional dominance by South Africa, in issues as varied as investment flows and intellectual capital, are often voiced (for example, Kibble et al, 1995, p 50; and Venter, 1996, pp 139–140).

At the same time, however, it is beyond doubt that SADC actively coveted South African membership before that country's ascension in 1994; moreover, its members would not want the country to leave the grouping now. It appears, therefore, that the regional attitude towards South Africa is a case of 'can't live with it, but can't live without it!' As such, it mirrors our propositions about the influence of hegemony, advanced in Chapter 2. Nevertheless, in spite of this ambiguity, we are fairly confident that the prospects for any kind of regional cooperation will be affected by the ways in which it impacts South Africa's relations with its neighbours.

Homogeneity

How homogeneous is the southern African region? On the one hand, supporters of SADC – particularly in its recent feud with COMESA – have been arguing that the similarities among the countries justify the strengthening of SADC (usually at the expense of COMESA). On the other hand, Blumenfeld (1991, p 138) notes that SADCC countries had 'widely diverse production structures, resource endowments, land ownership patterns, development priorities, institutional affiliations and resource allocation systems'. To try to uncover the extent of the similarities and differences within the region, any of a long list of national and regional characteristics could be examined.

Without undertaking such a systematic inventory, a few observations can still be made. Economically, the region encompasses countries characterized by the World Bank (1996a) as 'middle income countries' (specifically, Angola, Botswana and South Africa) and countries that are among the poorest in the world (Mozambique and Tanzania).[6] Differences in the countries' respective human development indices tell a similar story.[7] A few countries, moreover, have economies that are relatively industrialized (such as Angola, Botswana, Lesotho, South Africa, Swaziland and Zimbabwe), while others are primarily agriculturally based (Malawi, Mozambique and Tanzania). Socially, meanwhile, because the vast majority share a similar colonial history, English is the region's predominant language (Portuguese in Angola and Mozambique, and French in the Democratic Republic of the Congo are the notable exceptions); Christianity is also the predominant religion in most of these countries (CIA, 1996).[8]

Based on such observations, we can make some tentative conclusions. Firstly, a range of so-called social indicators would suggest that there is an English-speaking, largely Christian core to the region, which is made up of South Africa, Lesotho, Swaziland, Namibia, Botswana, Zimbabwe and Zambia. However, economic indictors, such as those mentioned above, suggest that the region as a whole is not particular homogeneous. Even when we consider only the seven countries in this 'core', convergence does not seem to occur: Botswana's per capita GDP is more than five times that of Zambia's (UNDP, 1997). Finally, a consideration of the respective countries' policy-orientation (membership of international fora, etc) suggests that many of the countries look outwards to the same organizations (for example, the World Trade Organization); but then again, so too do many in the developing world as a whole.

Like virtually all regions, southern Africa has both its similarities and differences. The particular pattern of those similarities and differences, however, might prove important in assessing the prospects for successfully implementing different regional arrangements.

Orientation

Regional Trade

As was argued in Chapter 2, orientation can be examined at many different levels. In the case of southern Africa, we begin by looking at international trade. Intra-SADC trade is not particularly high: before South Africa's entry to SADC, it stood at about 4 per cent of all trade conducted by the organization's then ten countries. In this way, SADC had closely followed the general African trend of trading more with those outside than with those inside the continent.

This phenomenon is, of course, well explained by historical factors. Aly (1994, p 43) argues that the 'former colonial metropoles monopolized the trade of their colonies and structured their economies so that the colonies would specialize in the production of primary commodities'. This colonial heritage manifests itself in all kinds of linkages in southern Africa. Weeks (1996, p 108), for example, reports that with 'few exceptions, the major arteries of transport in southern Africa reflect the historical integration of the region's trade with former colonial powers', and Hawkins (1994, p 104) declares that '[t]radition dies hard in African business. Thirty years after independence, many, possibly most, African countries still do most of their foreign trade with their former colonial powers.' Gambari (1991, p 13) argues that this orientation may be difficult to redirect: 'to those Southern countries that have invested in industrial growth aimed at exporting to the West, the idea of a de-linkage of their economies from those of the North is clearly unattractive'.

When South Africa is brought into the picture, intraregional trade flows grow considerably (see Table 3.1 for details, remembering the caution with which trade figures should be viewed).[9] It is worth recognizing, therefore, that much of southern African trade revolves around the Southern African Customs Union (SACU, which consists of Botswana, Lesotho, Namibia, South Africa and Swaziland). Table 3.1 shows that trade flows between SACU and each of Zimbabwe, Mozambique, the Democratic Republic of the Congo and Zambia are the largest in the region. It is also worth noting that there are significant imbalances in regional trade: the SACU countries export much more to the rest of the region than they import from the same.[10]

Though most of the economic orientation that exists revolves around South Africa, there are still other intraregional links. For example, Zambia and Zimbabwe have established a new interstate company called Railway Properties. It is intended to administer the non-divisible assets of the former Rhodesia Railways (EIU, 1996c, p 22). The Beira Corridor, between Mozambique and Zimbabwe, has also experienced a considerable increase in activity during the 1990s. Therefore, there is evidence to suggest that some kind of internal orientation in economic activities is developing, at least along the region's spine.

Table 3.1 *Estimated regional trade in southern Africa, 1996 (millions of US$)*

	Angola	DRC	Malawi	Mozam- bique	SACU	Tanz- ania	Zambia	Zimb- abwe
Exporters								
Angola	—	..	..	..	..	..	0	0.4
DRC	..	—	..	..	137	1.34	0.8	0
Malawi	..	4.87	—	4.64	69	4.31	5.44	5.2
Mozambique	..	..	5.47	—	37	0.93	0.2	6.9
SACU	..	176.0	238.6	597	—	154.03	323.2	1054.5
Tanzania	..	1.9	3.86	2.36	5.08	—	3.6	0.8
Zambia	0	51.6	5.08	0.32	37	12.64	—	8.6
Zimbabwe	15.4	29.0	113.8	96.46	360	39.97	74.8	—

Note: .. indicates data unavailable.
Source: IMF, 1997

Political Orientation

Southern Africa appears to have some kind of advantage with regard to political orientation: formal mechanisms for cooperation have existed for over 15 years in the form of SADC(C), which in turn built upon the relationships forged among the frontline states during the liberation struggles in the region. Indeed, the relative success enjoyed by SADC(C) (as argued in an earlier part of this chapter) appears to bode well for future regional cooperation that might require enhanced political links.

At the same time, however, the reader must recognize that not all of the region's relationships have been harmonious. First of all, there is some resentment between different countries' leaders, often because of what has happened in the past. For instance, the ultimately victorious liberation movement in any given country was not always supported by all neighbouring leaders! Moreover, many economic relations (which have often been conducted by political masters) have been acrimonious. Note, in particular, the generally abrasive atmosphere that has accompanied many bilateral trade negotiations in the region – particularly those involving South Africa; see above and Kapata (1997, p 23). Therefore, just because there has been political (or indeed, economic) orientation, do not assume that this has necessarily laid a sound foundation for future cooperative initiatives.[11]

Sociocultural Orientation

Østergaard (1993, 42) argues that there will not be sufficient contact at the grassroots level in Africa to sustain the 'development integration model'. He maintains that successful functionalism presupposes something that was present in the European case, but that is not in evidence in much of Africa – namely, 'modern associational pluralism, i.e., functionally specific, universalistic, achievement-orientated groups, such as interest groups and political parties' (Østergaard, 1993, p 41). This does not bode well for the presence of a regional civil society in any part of the African continent.

Datta (1989, p 97) follows a similar theme, focusing particularly upon southern Africa:

> Many non-governmental regional associations in Southern Africa have functioned only sporadically. Accelerated spurts of activities mark the periods on the eve of, during, and immediately after a conference. At other times associations lie dormant. Formal leaders are either too busy with their 'normal' duties or fail to attach adequate importance to regional work which thus suffers in comparison with the leadership function within the country.

In a more recent comment, Gumende (1996a, p 6) effectively illustrates these general assertions with a specific example:

> What can make the process difficult is the fact that efforts to consolidate a regional identity, through the community building process, have lagged far behind the development of instruments of politics, which is driven by politics. One example is the decision to establish 'SADC counters' at every airport in the region, a symbolic step towards facilitating the movement of people. Although the counters have been erected in most international airports around the region, they are rarely manned. The explanation for the bureaucrats' reluctance is that SADC is yet to prove its relevance to the people on the street.

Indeed, many have argued that governments and civil society in southern Africa have not interacted sufficiently; as a result, SADC is often portrayed as a heavily top-down organization.[12]

On the other hand, Peter Vale (1996, pp 384–386) suggests that, in spite of southern African states' efforts to reinforce traditional notions of national security, new (sub)regional maps are being drawn by different parts of civil society – a 'vibrant spirit of regionalism-from-below', as he calls it (1996, p 383). With South Africa's re-entry into international society, all kinds of interactions between its citizens and others in the region have certainly increased. However, formally measuring the extent of these interactions is, of course, difficult to do. Nevertheless, what is clear is that the degree of regional orientation, not only among all of the region's countries, but within various subsets, may affect the prospects for successfully implementing any particular regional mitigation option.

External Factors

Transnational Corporations and Investors

Although there are relatively few data about the impact of transnational corporations (TNCs) upon African cooperation schemes (Kisanga, 1991, p 41; and UNCTAD, 1995, pp 51–52), there is little doubt that (externally based) TNCs have exercised considerable influence upon the prospects for African development over the years (for example, Cantwell, 1991).[13] Moreover, given that many leaders in the region are, in contrast to some earlier attitudes, eager to increase the level of TNC activity in their countries, this level of influence may be set to increase (for example, Rowlands, forthcoming).[14]

In southern Africa, about three-quarters of all net foreign direct investment (FDI) between 1991 and 1995 flowed into either Angola (primarily attracted to that country's petroleum industry) or South Africa (UNCTAD, 1997, Annex Table 2). Of this, much comes from the United Kingdom and the United States, which would suggest that their potential attitude to any scheme for regional cooperation could be important. What, in particular, are they looking for? A recent UNCTAD study suggests that potential FDI agents in Africa would look favourably on two particular characteristics. One is movement towards greater regional integration, since this would serve to create larger markets (UNCTAD, 1995, p 93). A recent report from the World Economic Forum (1997) confirms this for the narrower case of southern Africa. (We should recall, however, that already operating TNCs may have reasons to oppose greater regional economic integration; see the discussion in Chapter 2.) The second characteristic is debt relief:

> The continuing debt problem prolongs balance-of-payments difficulties which, in turn, make it difficult to ease access to foreign exchange and profit-remittance regulations — an indispensable ingredient of any good investment climate, as investors may fear that reforms may be unsustainable and taxation too high if governments have to allocate substantial amounts to debt servicing. In addition, debt-servicing deprives Africa of badly needed domestic investment resources that could be used to improve infrastructure and encourage investment, thus enhancing the prospects for economic growth — one of the single most important FDI determinants. (UNCTAD, 1995, p 94)

Therefore, schemes which would increase dramatically the debt burden of southern African countries could well be resisted by TNCs and hence, by extension, many in southern Africa itself.[15]

Turning to the other key element of external investment, let us analyse portfolio investment. Table 3.2 provides some information about the stock markets in the region that operated during 1994. (Since that time, exchanges in Zambia and Malawi have opened; plans for Tanzanian and Mozambican ones are also well developed.) Although African markets have not attracted as many resources (or as much attention) as the other emerging markets in Asia and Latin America, it is still the case that substantial sums of money are involved – particularly in South Africa and, to a lesser extent (though still important in terms of share of GDP), Zimbabwe.[16]

Moreover, just as with FDI, southern African countries have shown a strong interest in encouraging foreign activity in their stock markets. Many have, during the past few years, relaxed or eliminated rules governing the extent to which non-nationals can buy and sell shares on the market. Therefore, we can again conclude that, all else being equal, the regional mitigation proposals that will gain the support of the markets are those that will have a better chance of being implemented.[17]

The next question is clearly: 'Who are these external investors?' The answer is not as evident as it was for TNCs' FDI. Instead, we can speculate that those countries with stock markets with the greatest capitalization values and the strongest culture of share ownership (the United States and the United Kingdom) will be most active in this part of the world.[18] Indeed, there are a number of funds for private investors, which focus on southern Africa, based in these countries.[19] Moreover, major pension funds have specialists studying the region.

How can we determine the attitude of the market in advance? If we knew that, then we would probably be very rich indeed. No particular correlation, apart from a favourable business climate, can be identified. Instead, we can speculate that the ways in which those involved in the market (for example, pension investors in Johannesburg and fund managers in London and New York City) think the market will respond to particular regional schemes may be enough to make their predictions come true. As such, the attitude of market investors to the different regional mitigation options that we develop could be influential.

Table 3.2 *Size of stock markets in southern African countries, end of 1994*

Country	Market capitalization (millions of US$)	Trading value (millions of US$)	Number of listed companies	Turnover ratio
South Africa	225,718	15,954	640	8.5
Namibia	9,574	18.4	14	0.3
Zimbabwe	1,828	176	64	11.5
Botswana	377	31.1	11	8.2
Swaziland	338	2	4	0.6

Source: IFC, 1995

Table 3.3 *ODA received in 1995, as a percentage of 1994 GNP*

Country	Percentage	Country	Percentage
Angola	7 (est)	Namibia	6.4
Botswana	2.3	South Africa	< 1 (est)
DRC	1 (est)	Swaziland	5.4
Lesotho	8.4	Tanzania	25 (est)
Malawi	34.1	Zambia	62.9
Mozambique	90.4	Zimbabwe	9.6

Source: UNDP, 1997, pp 190–191

Foreign Governments

In order to estimate the potential influence of foreign governments, one place to start is the figures for relative aid dependence – that is, the ratio of official development assistance to the recipient country's total GDP. Figures for the year 1994 are provided in Table 3.3. From this, we see that aid dependence varies widely in the region: from a low of less than 1 per cent in South Africa to a high of over 90 per cent in Mozambique. What is clear is that – if there is any truth in the rather crude hypothesis that influence can be bought with money – external governments may be able to exercise influence over some southern African countries (their past influence in SADC(C) is rarely questioned). Whether this hypothesis is true or not will not be fully explored here. Instead, the figures suggest that external donors may have some interest in what is going on in the region.

Having speculated that foreign governments could be influential, the question that is obvious is: 'Which one or ones?' To answer this, we turn to Table 3.4, where the total amount of ODA (net) given to 12 southern African countries is listed, broken down by donor government. From this, we see that those countries that have large absolute aid programmes in general – for example, Germany, the United States and Japan – are similarly active in the southern African region. We also see that one large country – France – is not particularly active. This, of course, comes as little surprise, given that there is only one former French colony, the DRC, among these 12. Finally, some countries with relatively small assistance programmes (in absolute terms) appear to have much of their activity concentrated in this part of the world – namely, Norway, Sweden and Denmark. Portugal also falls into this latter category; the country gave over one third of its ODA budget of US$271 million to the region (OECD, 1997; and UNDP, 1997). Not surprisingly, its interest is almost exclusively concentrated in its two former colonies in the region: Mozambique and Angola.

Though these numbers suggest that the impact of large OECD countries may be important (at least in some parts of the region), others argue that the influence of external governments may be minimal: 'In the final analysis, however, the capacity to build a "balance of prosperity" in Southern Africa

Table 3.4 *Sources of net ODA to southern African countries, 1995*

Country	ODA to southern African countries (millions of US$)	ODA to the region, as percentage of total ODA
Germany	488.0	6.5
United States	402.0	5.5
Japan	397.2	2.7
United Kingdom	311.6	9.9
Netherlands	308.5	9.6
Sweden	236.3	13.9
Norway	234.2	18.8
Denmark	184.7	11.4
France	157.5	1.9

Sources: OECD, 1997, and UNDP, 1997

will be determined within the region itself' (Blumenfeld, 1991, pp 163–164). Nevertheless, if there is any influence at all to be exerted, it will probably be done by the United States, the United Kingdom, Germany, Japan, The Netherlands and the Nordic countries.[20]

Intergovernmental Organizations

Questions about the relative importance of different international organizations (IOs) in southern Africa are usually answered with reference to two such entities: the World Bank and the European Union. Recent developments in the structure of the world economic and financial systems suggest that to these might be added a third candidate – namely, the World Trade Organization (WTO). Although Asante (1995, p 576) argues that the concept of African integration 'has been accepted by most of the international institutions as fundamentally important for African development', it is important to try to ascertain whether there are any particular nuances within this general position. Therefore, let us briefly identify the role of each in the region and highlight any evident attitudes towards regional cooperation in southern Africa.

The World Bank's operations are, of course, well documented. Active in every country in the region, in 1995 the bank sent US$632.2 million to the region through the International Development Association (IDA) (OECD, 1997). Weeks (1996), for one, argues that the World Bank has changed its tune and that it now lends support to efforts for African regionalism. Though still preferring multilateralism more broadly for economic and trade issues, the World Bank appears more than willing to advance regional options in response to certain challenges, including many environmental ones (World Bank, 1996b).

The European Union (EU) – more than the sum of its very important parts – might also be crucial. For one, the Commission of the European

Communities, as an entity, gave US$583.5 million in ODA to the region's countries in 1995 (OECD, 1997). Politically, meanwhile, the EU, in September 1994, agreed a Ministerial Declaration with southern African countries. In this, they, among other things, '[are d]esirous of promoting national reconstruction, regional cooperation and integration in Southern Africa; [and r]ecognising that the Southern African region has a substantial development potential through closer co-operation'. Moreover, one of the objectives of the agreement is that of 'promoting co-operation in trade with and in the Southern African region, in order to enhance its economic development'. The EU offered to share its experience in the field of regional integration with SADC as well (EU, 1994). How the present negotiations on a free trade arrangement between South Africa and the EU proceed, as well as the renegotiation of the Lomé Convention, may determine European influence in the region's future (for example, Keet, 1996; and Kibble et al, 1995, p 42).[21]

Finally, how the WTO responds to trade initiatives at the regional level may be crucial in the future. Indeed, even before we get to the stage of a regional trade arrangement, the bilateral agreements that are being discussed (for example, South Africa with Zimbabwe, which has a long history, and South Africa with Zambia, which is more recent) will continue to come under scrutiny. South Africa has been classified as a developed country by WTO. Consequently, some think that the terms of the bilateral agreement with Zimbabwe may have to extend to all developing countries, under WTO rules. Alternatively, however, this agreement may get a waiver: it was in existence before the WTO's formation in 1993.[22]

With respect to the issue of a potential regional grouping, however, some believe that the WTO will take a much tougher stance on regional agreements.

> *Experts say that the formation of trading blocs is likely to slow down in the near future. This is partly because there is a global concern that powerful regions can pose a threat to the multilateral trading system and trading blocs will now be thoroughly scrutinised by the WTO. Trading blocs are required to formalise their existence with the organisation and [Sheila] Page [of the Overseas Development Institute in London] feels that the Regions Review Committee, the watchdog that monitors compliance with the trade rules of the WTO 'is likely to take a much more rigorous approach to the details of regional agreements than in the past'.* (Quoted in Gumende, 1996c, p 24)

Indeed, this is further complicated by the fact that a SADC free trade area would, as presently conceived, consist of both developing and developed states (Gibb, 1996, p 11). In summary, the key point is simply that the WTO may influence any situation, and it will surely take an interest in the general proposals for a free trade area that SADC is considering.

SUMMARY

This chapter had two purposes.[23] The first was to examine the experience of the region's two most visible international organizations: SADC(C) and PTA/COMESA. Explanations for the relative successes and failures of these two bodies were viewed against the broader conceptual framework that was developed in Chapter 2. Though no unequivocal conclusions emerged – indeed, some of the reported explanations were themselves inconsistent – many of the more general ideas about what discourages and encourages regional cooperation seem to have also arisen in the case of southern Africa, particularly the importance of sovereignty (and associated national interests) and external forces.

The second purpose was to review the contemporary landscape in southern Africa. To achieve this, the chapter examined the agents (either individually or within bilateral or multilateral relations) and structures (both formal and informal) that could potentially affect implementing plans for regional cooperation. This mapping of the region was directed by the findings from Chapter 2. The messages from this section were also mixed: both optimists and pessimists could undoubtedly find support for their particular predictions concerning the future course of regional cooperation in southern Africa. Nevertheless, two specific findings merit reiteration: firstly, South Africa's role in most future arrangements will be crucial; and secondly, proposals that encompass only particular subsets of countries or sectors in the region may have more chance of success than universal proposals (moreover, which countries are included in each subset may well change from sector to sector). This second finding reinforces other individuals' conclusions (for example, AfDB, 1993a, p vi; Maasdorp, 1992, p 45; and quotations in Haarlov, 1997, p 22). All findings from this chapter, however, will be readily available when we embark upon our specific examination of regional mitigation proposals.

On its own, this chapter can be taken as an overview and analysis of efforts to promote regional cooperation in southern Africa. As part of the wider scope of this book, it is intended to suggest how both the present and past might help us to assess the prospects for implementing regional mitigation plans. Although we, of course, accept that the present and the past do not inevitably dictate the future, important factors may have been highlighted. As a result, we will keep in mind the findings from this investigation when we examine climate change cooperation in the rest of this book. We begin this more specific investigation in the subsequent appendix, where the history and present state of regional power sharing are reviewed.

ENDNOTES

1   The true inspiration for SADCC remains the subject of some debate. Anglin (1983, p 685), for one, points to the possible role of Zimbabwean President Robert Mugabe, 'who claimed paternity on behalf of both the Pan-African Freedom Movement of Eastern, Central and Southern Africa (1958–64) and the Conference of East and Central African States (1966–74)' as well as the European Economic Community, 'or at least certain of its officials'. For their part, Leys and Tostensen (1982, p 52) identify both internal actors (for example, leaders of the frontline states) and external actors (diplomatic initiatives on the part of African and Western states) as playing particularly formative roles.

2   Moreover, many argue that whatever development occurred only served to exacerbate inequalities among SADCC members by favouring Zimbabwe (Gibb, 1994, p 225; Hawkins, 1992, p 116).

3   Mwase (1994, p 33) argues that the PTA agreement ended up 'more or less copying the EEC Treaties with clauses from the Andean Pact thrown in'.

4   The 'uneven distribution of benefits' is also advanced by Foroutan (1993, p 249) as a reason for failure. In 1990, for example, 'Kenya and Zimbabwe accounted for respectively 32 and 27 percent of intra-PTA exports while absorbing only 15 and 7 percent of total imports' (Foroutan, 1993, p 258).

5   Consider not only the discussion about orientation, but also the fact that many southern African businesspeople are encouraging a regional outlook (see, for example, World Economic Forum, 1997).

6   Economic differences *within* countries would be even more pronounced.

7   In 1997, five of the 12 countries in the region were each classified as 'medium human development' while the seven others were classified as 'low human development' (UNDP, 1997; see the exact figures in Table 1.2).

8   Mozambique's recent joining of the Commonwealth, however, suggests that there might be an even greater degree of 'social homogeneity' present.

9   Smuggling and poor reporting generally are only two of the challenges that make the publication of figures that accurately reflect international trade flows difficult to achieve.

10 Of course, the trade among the five members of SACU is not shown by these figures. There are significant economic links among them, with considerable trade as well – and most of it probably involves South Africa.

11 A consideration of political orientation provides an appropriate time to note the following observation, from the Joint SADC/PTA study of regional prospects:

    We are deeply conscious of the fact that Tanzania is historically and politically an integral member of SADC; indeed for many, it is inconceivable to speak of SADC without Tanzania. Equally, however, Tanzania is an integral part of East Africa. The decision as to whether Tanzania remains in SADC, becomes a member of the proposed East African Community or enjoys membership of both SADC and EAC, is a matter between Tanzania and the member states of these groupings. (Mandaza et al, 1994, p 44)

12 See, for example, Katerere et al (nd) and Kibble et al (1995, p 61).

13 Here we are primarily concentrating upon TNCs that have their headquarters outside of southern Africa. Nevertheless, many of the ideas developed will also apply to South African TNCs, operating in other countries of the region. These latter entities may well prove to be crucial.

14 An UNCTAD report (1995, p 37), for example, reveals that foreign direct investment 'is welcomed and, in fact, actively sought by all African countries'.

15 The impacts of TNC FDI, and the relative merits of the same, are not being explicitly investigated here. This is not, however, meant to suggest that they are not subjects worthy of study. Simply, because there is a consensus that attracting FDI to southern Africa should be encouraged, proposals that do not interfere with this objective are more likely to be supported.

16 The relative size of the Namibian exchange is not a good indicator of its importance: it is comprised primarily of South African companies (which have an additional listing in Windhoek), and it has a low level of activity (as indicated by its low turnover value).

17 As was the case with FDI, this statement is not meant to imply an uncritical acceptance of stock market expansion. Instead, it is cited in recognition that statements like 'what is good for the Johannesburg Stock Exchange is good for South Africa' have broad support. As such, a regional mitigation scheme that attracts the ire of the markets could well encounter large implementation obstacles. Moreover, it is also worth noting that much of the local ownership of shares is by institutional investors – pension funds and the like (Jefferis, 1995). People employed in the major southern African organizations, both private and public, will not want to see their retirement nest egg disappear. Accordingly, we can speculate that they will, all else being equal, also want a strong stock market.

18 The Morgan Stanley Capital International World Equity Index allocates a 44 per cent share to the United States, 15 per cent to Japan and 10 per cent to the United Kingdom ('Our Portfolio Poll', 1997, p 80).

19 Save and Prosper's Southern Africa Fund, for example, aims to 'achieve long term capital growth from investment primarily in the securities of companies quoted or trading in the countries of Southern Africa, or in closed-ended vehicles which invest primarily in such securities' (Save and Prosper, 1998).

20 Kibble et al (1995, p 52) identify the United Kingdom, Sweden and Finland as being the only EU members with 'a traditional interest in Southern Africa'. Moreover, recent writings in *The Economist*, for one, suggest that the United States is overcoming its traditional 'Afrophobia' ('America Loses its Afrophobia', 1997).

21 Keet (1996) suggests that the EU may want to integrate South Africa more closely into the neoliberal world. This may have implications for South Africa's relations with its neighbours: without Lomé membership for South Africa, there may be little encouragement for regional cooperation in manufacturing, since regional goods would presumably not be given preferential access to the European marketplace.

22 This resembles the way that SACU has been accepted as an exception to the rule. However, what might develop if SACU were to take on new members is unclear.

23 Though not explicitly examined (because of their more obtuse nature), interest in both linkages and capacity (see Chapter 2) will also be mentioned throughout this study.

Appendix	Experience with Regional Power Sharing
	R S Maya

SOME EXPERIENCES IN REGIONAL ELECTRICITY COOPERATION

Zambia–Zimbabwe's Early Experiences

Electricity interconnections, such as those developed during the mid 1950s to supply the copper mines in what is now northern Zambia, and industry in what is now Zimbabwe, were established for a clear purpose and under amicable political conditions (adapted from Maya, 1982). With respect to the former, production bottlenecks were forming in both regions, primarily because of a shortage of power. With respect to the latter, Zambia and Zimbabwe were part of a single economic unit – namely, the Federation of Rhodesia and Nyasaland. To offset this electricity deficit, the affected mining companies combined resources to form the Rhodesia–Congo Border Power Corporation, Rhopower, in order to draw electricity from what is now the southern part of the Democratic Republic of the Congo and to transport it to the copper belt and on to Zimbabwe. This scheme was one of southern Africa's earliest regional interconnectors. While the quantity of energy transferred was small, it is significant that Rhopower was a wholly private initiative: one that arose within highly enabling political and economic environments.

The second major regional scheme, the Kariba Hydroelectric Scheme, was much larger. Involving over 1200 megawatts (MW), it created what was then the largest man-made lake in the world. The Kariba project, however, was realized under different political circumstances. Ideas about building the Kariba Dam can be traced as far back as 1912, when a local district administrator proposed a dam for agricultural development. Initial surveys by two irrigation engineers were conducted in 1914. In 1927, another private party approached the government, but the idea was dismissed as impractical by the government's own engineers.

The concept of Kariba as an energy project was first proposed in 1941, when a statutory body responsible for electricity supply pressured the government for funds to study the Zambezi River for hydroelectric purposes. The results were conclusive, and construction began in 1955 with initial transmission set for 1961. Funding came from the federal government with a loan from the World Bank. The Kariba project was viewed as positive for many reasons, particularly in terms of securing adequate energy supply and encouraging the development of secondary industry in Zambia and Zimbabwe. A team of British industrial and process engineering consultants, in a report entitled 'The Development of Manufacturing Industry Within the Federation', argued strongly that the Kariba system would enable development in the electric reduction of chrome to high-grade ferrochrome. (At least 75 per cent of the world's reserves of high-grade ferrochrome are in Zimbabwe.) The consultants also made special note of the manufacturing and agricultural processing industries as sectors whose development would be boosted by the fixed-cost power from the Kariba project.

Despite its significance and its perceived contribution to the energy resource base of the two countries, the Kariba project was hardly viewed as the panacea to the power problems that the region faced. Nor was it thought to be the countries' ultimate power base – at least not at the time of commissioning, when the Kariba facility had only one power house on the South Bank. At the time, the federation envisaged a 9 per cent annual increase in electricity demand, thus effectively doubling every eight years. With industrial expansion in mind, and with such plans as saving wood by adopting the use of electricity in tea drying, and a boom in the high energy industries of fertilizers and explosives, a major power shortfall was expected by 1974; the expansion of Kariba was therefore seen as a timely investment.

Even with such expansion, the Kariba facility was not regarded as capable of meeting the nations' energy needs for long on its own, since another power shortfall was expected in 1978. The solution this time was not to expand or retrofit the Kariba Dam. Rather, a giant thermal power plant was planned for construction at Zimbabwe's coal base at Hwange. There were two main reasons for moving away from the hydropotential of the Zambezi. Firstly, it would provide insurance for power supply in the event of droughts. And secondly, it would provide an outlet for lower-grade coal produced at Wankie Collieries, where the main output was industrial and coking coal.

It is clear that the establishment of regional hydroelectric power followed no clear formula. Rather, there was a mix of political interests and economic needs which were quite persuasive. An institutional mechanism in the form of the Federal Power Board, which changed into the Central African Power Corporation (CAPCO) and later into the Zambezi River Authority (which is how it presently exists), ensured effective technical management of the Zambezi waters for energy purposes. Nevertheless, the advent of large-scale thermal systems reveals that there was official apprehension about overreliance on the Zambezi. This led to the perception that hedging against drought and loss of supply due to the turbulent geopolitics prevailing at that time was clearly necessary.

Regional Interconnectors

One of the region's most important interconnectors links the Zambian system and the system of the Democratic Republic of the Congo (DRC). Rated at 220kV (kilovolts), it was meant to secure electricity supply to the two countries' mining areas. After completion, it was envisaged that electricity would flow either way, balancing out by the end of the financial year; as such, no payments from one country to the other would be necessary. But in 1989, a major fire at Zambia's Kafue Gorge power station left the country with a deficit which could only be met by importing electricity from the DRC. To this end, a high-voltage electricity supply agreement between the Zambian utility, ZESCO (Zambia Electricity Supply Corporation Limited), and the Zaïrian one, SNEL (Société Nationale d'Electricité), was signed.

During 1992 and 1993, ZESCO again had to import electricity on a commercial basis, because of the unprecedented drought in the subregion. From 1993 to 1994, the company continued to import electricity from Zaïre, though this time as part of a tripartite agreement for wheeling electricity from Zaïre to Zimbabwe through the ZESCO network. The agreement was necessitated by the fact that Zimbabwe could not import adequate supplies from Zambia, its traditional import source, because of a regional drought which limited power production in Zambia on both the Kafue River and the Zambezi River. Imports from Zaïre amounted to 120 MW and were wheeled through the Zambian system to the Kariba South Bank.

A second important interconnector in the region is the one known as the SADC Interconnector. This is the 220kV line linking Bulawayo in Zimbabwe and Francistown in Botswana. In the late 1980s, the rationale for building this line was the desire for Botswana, as a member of SADCC, to reduce its dependence upon South African supplies. The line was built under the auspices of SADC–TAU with CIDA (Canadian International Development Agency) financing the initiative. The effectiveness of this interconnector, however, was hampered by two events. The first was the change in regional geopolitics, with South Africa becoming an acceptable member of the southern African community. As a result, Botswana could more easily import from its neighbour to the south. And second, the line also encountered stability problems associated with a design error – more specifically, the lack of automatic generation control. Although an agreement for the supply of firm and surplus power was signed among Botswana Power Corporation (BPC), ZESCO and Zimbabwe Electricity Supply Authority (ZESA) after lengthy negotiations, it never reached its full potential because of stability problems. In reality, the line has been primarily used to supply surplus energy. Even this way, however, the benefits to Botswana could be significant; it can save up to 20 per cent of its generation costs by purchasing surplus power from Zambia and Zimbabwe when available.

The Southern African Power Pool

Power sector collaboration is the area in which SADC countries have made the most significant strides towards greater regional cooperation. In December 1995, an Inter-Utility Memorandum of Understanding was signed among nine of the then 12 SADC utilities and the power utility of Zaïre (then a non-SADC member). This created the Southern African Power Pool (SAPP).

The formation of SAPP was motivated by a desire to enable regional power transfers to level off regional supply imbalances, to increase regional security of supply, to smooth load curves, and to engender economies of scale in the supply base. Further benefits were envisaged, including increased revenue for the exporting country by opening up a ready market for excess generation capacity, lowering spinning reserve costs, and harnessing water resources which would otherwise flow unutilized to the oceans or evaporate from reservoirs. For the importing country, meanwhile, benefits would include avoiding the costs of unserved energy, reducing spinning reserve margin, deferring capital investment, and benefiting from sharing costs and risks through joint planning and investment.

SAPP institutions are already operational: working committees and regular meetings of utility chief executives exist to discuss policy and strategic issues. A SAPP communications system, which will be located in Zimbabwe, is also being established. Moreover, programmes are being planned to provide specialized training in the field of electric power. These may include training for operations under SAPP.

The region's present power sharing arrangements, which are based primarily on interconnections, provide a good network for sharing power to meet national shortfalls and to offset temporary deficits. However, interconnections for these reasons do not yield the maximum financial and environmental benefits that are possible for the region. In order to secure maximum benefits, the region would have to adopt and implement 'power pooling' as an operational strategy for installed power stations and transmission networks.

An advanced version of regional interconnections, power pooling is based on offsetting national deficits. It involves selecting power plants to run on the basis of, for example, cost or environmental benefits (whether local, national, regional or global, or some combination of the same). The series of plants from which this selection would be made are usually owned by one large utility or are located entirely within one country. Because of southern Africa's relatively small power base, however, power pooling should be considered among all of the region's countries. Therefore, given that the plants would be in different countries and therefore would be owned by different national utilities, the idea of power pooling would be much more complex to implement.

| Regional Electricity Demand and Supply: Developing the Baseline

Bothwell Batidzirai, Norbert Nziramasanga and Ian H Rowlands

INTRODUCTION

In this chapter, we establish the baseline for our subsequent investigation into mitigation options in the power sector. In other words, we present a scenario for the future development of the region's power sector, in the absence of substantial concern about global climate change. Temporally, we examine prospects up to, and including, the year 2050; while spatially, we focus upon the 12 mainland members of SADC.[1] At the end of this chapter, we estimate the region's electricity demand over the next half century and provide information on how that demand would be met (that is, supplied) in a business-as-usual scenario.

The chapter is divided into two main parts. The first uses information provided by the respective countries' power utilities to describe, as far as possible, the collective vision of the future for the region. This is, of course, a logical point of departure, since electricity planning is one of the utilities' primary responsibilities. Therefore, we consider each of our 12 countries individually, providing a brief profile of electricity demand and supply, both present and future.

Information from utilities, however, reveals only part of our desired picture. For one, each utility in the region has undertaken future planning to a different extent, and with varying degrees of openness. Therefore, we might find that the available information does not cover the entire time period that we are investigating: up to and including the year 2050. As a consequence, were we to rely solely upon the utilities, we might only end up with part of the information we require. Furthermore, the information that is provided might not necessarily be consistent. Though we certainly acknowledge that many of the region's utilities possess some of the most complete information on electricity planning, we should, nevertheless, still closely scrutinize their projections. Different utilities may have different projections – perhaps concerning import and export levels, or the commissioning of a large-scale project. We need to ensure that the regional scenario that we collate is internally consistent.

Consequently, in the second part of the chapter, we supplement the utilities' information with our own analysis. Thus, we develop a picture of the region's future electricity demand and supply. We express this in terms of maximum power and in terms of the source (which will therefore allow us to calculate greenhouse gas emissions in the next chapter).[2] This baseline will serve as the reference with which to compare our mitigation scenarios.

THE UTILITIES' PROFILE OF THE REGION

Angola

Most of Angola's available power system is hydrobased: in 1995, hydropower accounted for 200.8 megawatts (MW) of 326.0 MW (or 61.6 per cent) of total available capacity. The balance (which includes backup) was supplied by diesel stations (82.8 MW) and gas turbines (42.4 MW) (Eskom, 1996a, p 48). Figure 4.1 provides the capacity details for all of the countries in southern Africa.

Because the Angolan power system is not fully integrated – in addition to a large northern system, there is a central system, a southern system, as well as isolated power stations at four sites and private generating capacity at the oil refinery in Luanda – it is difficult to talk about peak demand for the system as a whole. Nevertheless, in 1995, the sum of peak demands for its various parts was 180.6 MW, well below available capacity (Eskom, 1996a, p 48).[3]

In 1995, total electricity produced by the country's national utility (therefore not including the 10 MW gas turbine owned and operated by private interests at the oil refinery) amounted to 1042 GWh, with 877 gigawatt-hours (GWh) (84 per cent) from hydro and the remaining 165 GWh (16 per cent) from the country's thermal sources (Eskom, 1996a, p 49).

Forecasting demand in Angola is extremely difficult. In the absence of information from the national utility, a recent report by two southern African energy non-governmental organizations (one regional and one South African) relied upon a 1992 study by the World Bank (SAD–ELEC and MEPC, 1996, pp 109–110).[4] Though this source is becoming increasingly dated, the authors of the SAD–ELEC/MEPC study clearly concluded that it warranted consideration. Because our investigation has failed to uncover any more recent (or more accurate) estimates, we use it for projecting future demand.

Assuming a middle road between the low and high scenarios developed in the World Bank study (which in itself may be optimistic, given the ongoing conflict in Angola), we calculate the following for future energy and peak demand: 2009.5 GWh in 2000 and 2795.5 GWh in 2005 for the former, and 378.5 MW in 2000 and 530 MW in 2005 for the latter

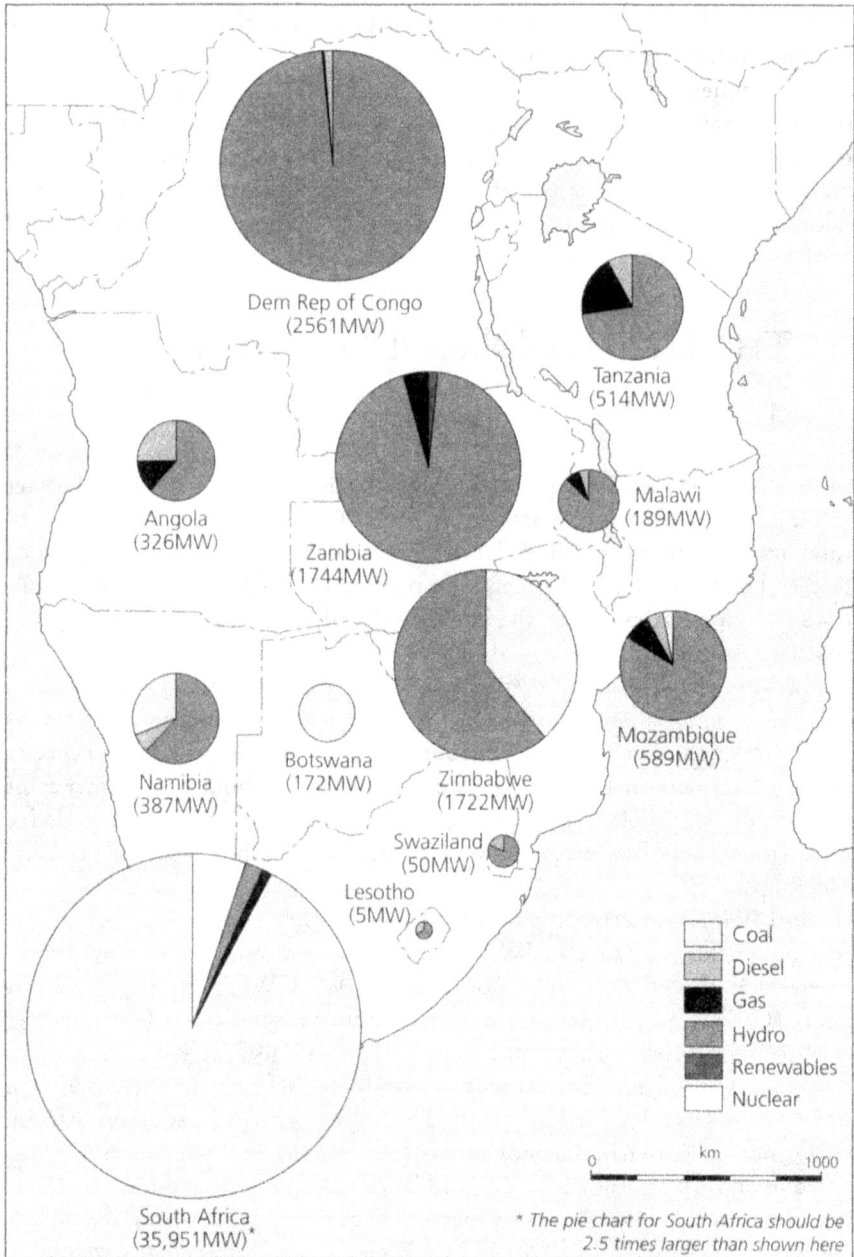

Dem Rep of Congo
(2561MW)

Tanzania
(514MW)

Angola
(326MW)

Zambia
(1744MW)

Malawi
(189MW)

Namibia
(387MW)

Botswana
(172MW)

Zimbabwe
(1722MW)

Mozambique
(589MW)

Swaziland
(50MW)

Lesotho
(5MW)

Coal
Diesel
Gas
Hydro
Renewables
Nuclear

km
0   1000

South Africa
(35,951MW)*

* The pie chart for South Africa should be
2.5 times larger than shown here

Figure 4.1 *Generating capacity in southern Africa, by resource and country, 1995*

(SAD–ELEC and MEPC, 1996, p 110). As a result, peak demand should overtake available capacity shortly before the end of the century.

Efforts to increase electricity supply in Angola have been hampered, and often interrupted completely, by the country's civil war. Nevertheless, they have concentrated upon the Capanda site, a facility which is scheduled to have four units, each rated at 130 MW (SAD–ELEC and MEPC, 1996, p 113). It is now reasonable to expect that the first two of these will come on line two years after the situation in Angola has stabilized, with the remaining two another year or two after that. For our baseline, therefore, we introduce these in 2000 and 2002, respectively. In addition to this, there could also be increases in supply as a result of rehabilitating existing power stations: 287 MW of Angola's existing system were reported to be out of order in 1995 (SAD–ELEC and MEPC, 1996, p 112). We assume that ten MW of this potential are brought on line every year between 1998 and 2004 (varying between hydro and thermal rehabilitation). After executing some basic interpolations on the demand side, we are left with a demand-and-supply profile until and including the year 2005. Full details are provided in Figures 4.2 and 4.3.

Botswana

With large reserves, coal dominates Botswana's indigenous generating capacity. During 1995/96, 805 GWh were supplied by the coal-fired power station at Morupule (which has a net maximum capacity of 118 MW) and 88 GWh were supplied by another coal-fired power station at Selebi–Phikwe (net maximum capacity of 54 MW). Supplementing this were imports from South Africa's utility, Eskom (383 GWh). Supply thus amounted to 1276 GWh (BPC, 1997).

The load forecast produced by the Botswana Power Corporation (BPC, the national utility) predicts an approximately 6.5 per cent annual increase in system peak load between 1995 (205 MW) and 2000 (284 MW). Further annual increases of 4.4 per cent during the subsequent ten years bring the peak load to 437 MW in 2010. Associated energy demands are 1659 GWh and 2419 GWh respectively (BPC, 1997).

To meet these anticipated increases in future demand, the government and BPC have committed themselves to more active use of the opportunities offered by regional supply arrangements (rather than the expansion of more costly indigenous coal-fired generation). At present, BPC anticipates that all of the increasing demand will be met by imports from South Africa. The quantity of such imports rises from 614 GWh in 1996 to 1756 GWh (or 68.7 per cent of total anticipated supply) in 2011 (BPC, 1997). The rest would be generated domestically by the Morupule power station. (Selebi–Phikwe was decommissioned in 1996.) This information, therefore, allows us to complete a demand-and-supply profile for Botswana until and including the year 2011. This is reported in Figures 4.2 and 4.3.

Peak demand (MW), log scale

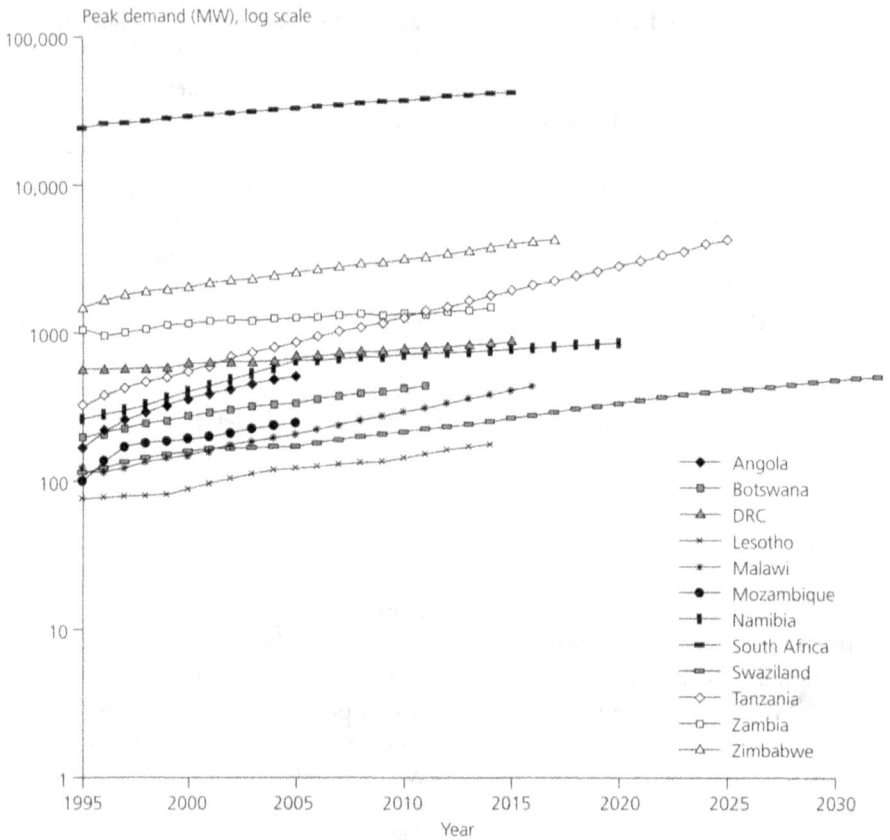

Figure 4.2 *Peak power demand in southern African countries, as projected by national utilities and similar sources*

Democratic Republic of the Congo (formerly Zaïre)

With a maximum capacity of 2560.8 MW (in 1994), the Democratic Republic of the Congo (DRC) represents the second largest supply country in our study. In the DRC, virtually all generation (98.5 per cent) is hydrobased, with one natural gas power station and 28 small diesel stations rounding out the supply profile. In 1993, these sources produced 5379 GWh of electricity. Surplus to the country's requirements, a substantial amount (876 GWh) was sold to Zimbabwe, with smaller amounts exported to Zambia, the Congo, Rwanda, Burundi and Angola (Eskom, 1996a, pp 70 and 71). Estimates for 1995 are that domestic consumption was about 3400 GWh, while exports (again mainly to Zambia, Zimbabwe and the Congo) totalled 1500 GWh. Peak demand was about 600 MW domestically and 250 MW on the export commitments (SAD–ELEC and MEPC, 1996, p 212).

In estimates attributed to both the national utility and SAD–ELEC, electricity demand in the country rises by a modest annual rate of 2 per cent

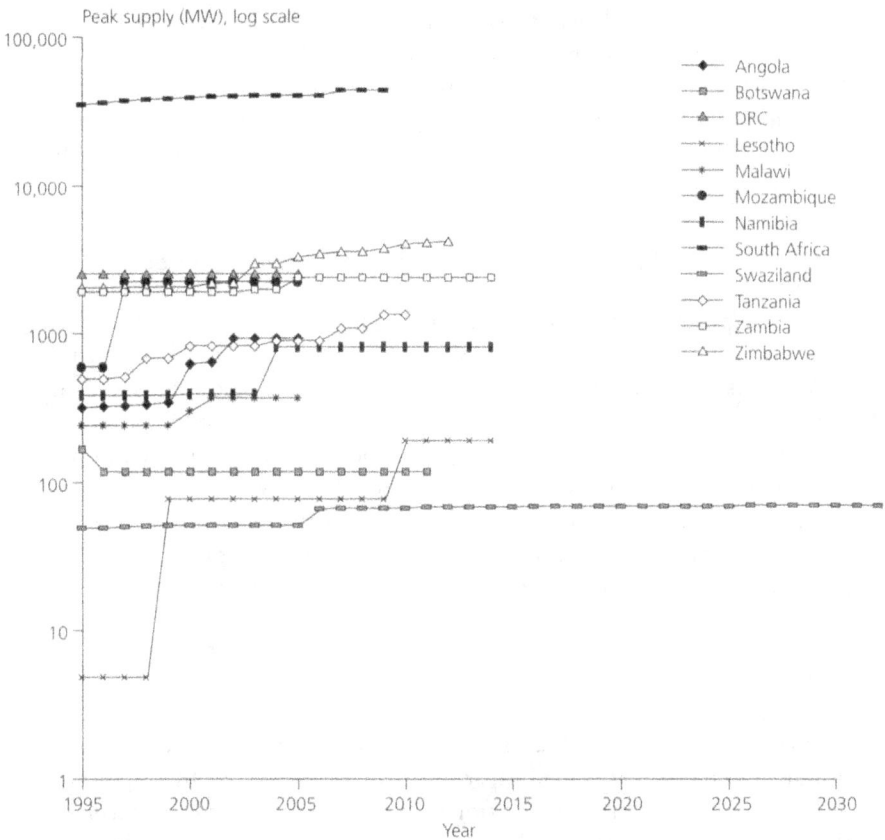

Figure 4.3 *Peak power supply in southern African countries, as projected by national utilities and similar sources*

during the period 1995 to 2015. Though the situation has certainly changed since those estimates were made (most significantly, the removal of Mobutu Sese Seko from power, and the installation of Laurent Kabila as president), uncertainty continues to characterize the country. For this reason, we adopt this estimate of demand increases for our tentative baseline until the year 2015 (and keep the supply profile constant to the year 2005). See Figures 4.2 and 4.3 for details.

Lesotho

In 1995/96, all of Lesotho's electricity was imported from the South African utility Eskom. Although there was a nominal capacity on the country's territory (about 3.3 MW of mini-hydro and 1.6 MW of diesel), none of it served to generate electricity for the national utility. Peak power demand in 1995/96, meanwhile, was 80.02 MW (Eskom, 1996a, p 56).

Gross electricity demand (as estimated in a medium alternative scenario developed for the Lesotho national utility in 1995) is expected to rise to 334.9 GWh in 1999, 503.5 GWh in 2004, 550.7 GWh in 2009 and 756.0 GWh in 2014. Associated peak demands for these four years are predicted to be 85 MW, 121 MW, 139 MW and 185 MW respectively (SAD–ELEC and MEPC, 1996, pp 135 and 134).

Central to meeting this increasing demand is the construction of the 72 MW Muela hydropower facility as part of the Lesotho Highlands Water project. Full commissioning is expected in 1998. Though this is the only confirmed project in the country's future generating plans, additional candidates include an extension of the Muela facility (where another 100 to 120 MW of installed capacity is possible); pumped storage capacity of up to 1000 MW has also been suggested (SAD–ELEC and MEPC, 1996, pp 138 and 139; see also Redeby et al, 1994). For our purposes, however, we assume that the first phase of the Muela facility will be fully exploited in the 1999 calendar year, while another 110 MW will be fully available in 2010. With this information, we construct a demand-and-supply profile for Lesotho until and including the year 2014 (see Figures 4.2 and 4.3).

Malawi

Electricity generation in Malawi is dominated by the country's hydropower resources. The total installed capacity of the country's six hydropower facilities is 219.1 MW, with an additional 15 MW provided by gas turbines and the final 9.6 MW from diesel generators. Thus, the ratio of hydropower to non-hydropower capacity is approximately nine to one (SAD–ELEC and MEPC, 1996, pp 145–146).

During 1994/95, total electricity sales reached 731.4 GWh, with a system peak load of 149.4 MW (Eskom, 1996a, p 58). Given that 54.5 MW of the power supply referred to in the paragraph above only came on line during 1995, Malawi's electricity demand and supply had, to that point, been virtually in balance. There was some border trade with Mozambique, but this was relatively small at about only one GWh (SAD–ELEC and MEPC, 1996, p 144).

The Electricity Supply Commission of Malawi, the national utility, has developed annual load forecasts to the year 2016 (Escom, 1997). Demand is expected to rise to over 450 MW by 2016. To meet this, there is a commitment to the Kaprichira hydropower project, which is expected to add 64 MW to capacity for the year 2000, along with a possible additional 64 MW the following year (SAD–ELEC and MEPC, 1996, p 148). We incorporate these figures into our supply baseline.

Beyond that, there are a number of possibilities for additional supply. For instance, discussions about interconnectors with Mozambique and Zambia suggest that imports could play a greater role in the future.

Furthermore, the country's hydropower potential has been estimated at 600 MW, implying that at least 250 MW remain unexploited. Australian interests, moreover, have investigated the possibility of developing a coal-fired power plant with a potential 300 MW capacity. (The plan, however, was to export the electricity generated; SAD–ELEC and MEPC, 1996, pp 141 and 150–51.) Finally, the construction of gas-turbine or wood-fired power stations (perhaps by independent power producers) is being considered (Escom, 1997). All such developments, however, are at a very preliminary stage. Consequently, our supply scenario does not incorporate any of them, as it extends until and including the year 2005, with the demand scenario until and including 2016. See Figures 4.2 and 4.3 for details.

Mozambique

Total electricity available for distribution in Mozambique (excluding system losses) amounted to 978 GWh in 1995. Over half of this (606 GWh) was imported from the South African utility Eskom. The vast majority of the remainder (336 GWh) was generated domestically by hydro sources. Small amounts of diesel-generated power, along with imports from Zimbabwe and Malawi, accounted for the remaining 36 GWh (Eskom, 1996a, p 62).

Based upon input from the Mozambican utility, SAD–ELEC has forecast future loads for the country. Estimates for 2000 are a demand of 1150 GWh and a system peak load of 200 MW. Corresponding figures for 2005 are 1430 GWh and 250 MW (SAD–ELEC and MEPC, 1996, p 155). These do not incorporate the possible implementation of a large industrial or mining project – for example, the construction of an aluminium smelter. Though recent interest in such projects suggests that one might be forthcoming, we are confident that any such development would be self-sufficient in terms of power; as part of any agreement, some dedicated source of power would probably be secured.[5]

Mozambique's supply profile changed dramatically in 1997, when repairs to the transmission lines from the 2075 MW-rated Cahora Bassa hydropower facility were completed. Consequently, exports to Zimbabwe and South Africa were secured. Although Mozambique has more than enough capacity to meet its own demands, there is still great potential for increased power generation in the country, by means of either hydropower or natural gas (particularly, in the case of the latter, at the Pande field). However, at this point, no firm plans for generating expansion have been finalized. Consequently, during the time horizon projected – that is, until the year 2005 – we do not add any new generating capacity (see Figures 4.2 and 4.3).

Namibia

Namibia has four sources of power on its territory: one coal-fired generator (with a maximum net capacity of 120 MW), one hydropower facility (240 MW) and two small diesel plants (combined capacity of 27 MW). Together, they provided 1259 GWh of electricity in 1994/95. This was supplemented by imports totalling 767 GWh (almost exclusively from Eskom, though a small quantity was supplied from the Zambian utility ZESCO). The system's peak demand was 277 MW (Eskom, 1996a, p 64).

The Ministry of Mines and Energy in Namibia has projected peak power demand in Namibia for the coming 25 years. It foresees steady annual increases of 10 per cent to 2005, with a 'levelling off' during the subsequent 15 years (presumably as the country enters the classic energy transition phase) (Ministry of Mines and Energy, 1997). Though specific details as to how this demand will be met remain unclear, the Namibian government recently declared that it wants to meet 75 per cent of its peak demand through internal sources by the year 2010 ('Energy White Paper Spells Out Focus', 1998).

Given this, and recent comments from government officials, the development of a large hydropower facility at Epupa on the Kunene River could well be part of the country's future supply plans. Prefeasibility studies have already been carried out, and it has been recommended that a facility with three generating units is constructed.[6] The total installed capacity would be 415 MW and average annual energy generation capability would be 1650 GWh (SAD–ELEC and MEPC, 1996, p 172). Therefore, we introduce this into the baseline from 2004. An additional 1400 MW of power could be harnessed from the Kunene River (which Namibia shares with Angola). Natural gas from the Kudu offshore field could also be used to generate electricity. We do not, however, put any additional supply options into our profile at this point. Consequently, we are left with a demand profile to the year 2020 and supply data to the year 2014, both of which we outline in Figures 4.2 and 4.3.

South Africa

South Africa dominates the southern African region on virtually every measure, electricity included. Coal is the main fuel for the country's electricity supply – not surprising given that South Africa is one of the world's largest coal producing countries and is also home to some of the world's largest reserves. More specifically, 92 per cent of the energy generated by the country's largest utility, Eskom, came from ten coal-fired power stations in 1995. Other contributors included nuclear (the continent's only nuclear power station provided 7 per cent), pumped storage (two plants contributing 0.8 per cent), and hydropower (0.3 per cent from two facilities).[7] Total effective capacity of all power stations (including those in reserve) was 35,951 MW. Most of the 164,834 GWh of electricity that were produced were used within the country's borders, with only 1.2 per cent sold internationally (to customers in Botswana, Mozambique, Namibia and Zimbabwe). An even smaller amount (172 GWh) was purchased from non-Eskom sources (Eskom, 1996a, p 4).

In its fifth Integrated Electricity Planning document (Eskom, 1996b), Eskom predicts an annual growth rate in its peak demand (after demand-side management initiatives) of 2.4 to 3.4 per cent until the turn of the century, approximately 2.5 per cent per year during the first decade of the new century, and decreasing to 2.1 per cent during the subsequent five years. In terms of absolute amounts, peak loads are expected to rise to 29,489 MW in 2000 and up to 42,332 MW in 2015 (compared with 25,133 MW in 1995). Corresponding figures for annual energy production are 194,107 GWh and 276,632 GWh respectively. To meet this demand, a number of initiatives have been proposed.

Firstly, an import agreement has been reached with Mozambique, so that Eskom imports 950 MW from Cahora Bassa, rising to 1450 MW after 2004. Eskom is also importing 350 MW from the Zambian utility ZESCO. Secondly, Eskom will recommission a number of generators currently on reserve. And thirdly, it will commission new capacity: by opening coal-fired units at Majuba (612 MW in each of 1996, 1997 and 1998, followed by 667 MW in each of 1999, 2000 and 2001); and by opening additional new power stations, both coal fired and pumped storage, between 2010 and 2015. For our baseline, we incorporate the import agreements and the new capacity at Majuba, as well as a new 3000 MW coal-fired power station in the year 2007 (to meet the country's rising demand), thus allowing us to construct a supply profile until the year 2009. Figures 4.2 and 4.3 present this, as well as the demand profile until the year 2015.

Swaziland

Swaziland's domestic contribution to meeting its own electricity demands is relatively modest. During 1994/95, the net maximum capacity of the country's system was 50 MW, with about four-fifths of that provided by hydropower and the remainder by diesel. Given that peak demand in that year was 117.5 MW, additional power had to be imported. Indeed, in terms of supply for the year as a whole, only 109.8 GWh were produced domestically (virtually all from hydro), while 597 GWh were purchased from the South African utility Eskom (Eskom, 1996a, p 66).

The government's Ministry of Natural Resources and Energy predicts peak load to increase steadily during the coming 35 years – averaging just under 4 per cent a year (Ministry of Natural Resources and Energy, 1997). To meet this demand, attention will focus upon transmission potential –

not only further linkage with the South African grid, but also interconnections with Mozambique. Nevertheless, there is also some interest in generation: investors have investigated the possibility of constructing a 200-MW coal-fired power station (though the economics did not end up being particularly favourable), and the exploitable potential of hydropower is assumed to be approximately 400 GWh (SAD–ELEC and MEPC, 1996, pp 194 and 192). However, the only additional indigenous source that we will include in our supply scenario to the year 2032 is the construction of a 15-MW hydropower facility at the Komati River Dam. At present, the sole function of this dam is agricultural (that is, irrigation); no power plans are currently being pursued (SAD–ELEC and MEPC, 1996, p 192). Let us assume, however, that its potential for generating electricity will attract some interest and come on line after 2005. This allows us, in Figures 4.2 and 4.3, to complete supply-and-demand profiles until the year 2032.

Tanzania

Electricity in Tanzania is predominantly hydrobased: of the 525 MW of available capacity reported in 1997, 375 MW came from five hydrofacilities, while an additional 112 MW came from a gas turbine and the final 38 MW were powered by diesel (Tanesco, 1997). With the gas turbine only being installed during 1995, most of the electricity available for distribution during that year came from the hydrofacilities (1539 GWh, or 86 per cent); virtually all of the remainder came from diesel generators (252 GWh) and only a small amount (11 GWh) came from imports (Eskom, 1996a, p 68).

Demand-and-supply projections have been made available by the national utility: Tanesco (Tanesco, 1997). Tanesco predicts a steady increase in demand during the next 30 years (about 7 or 8 per cent annually). To meet this, additional supply will be provided by the following facilities:

- 1998: new gas turbine (37.5 MW); and new diesel power by independent power producers (85 MW);
- 2000: new hydrodevelopment at the Lower Kihansi (180 MW), an interconnector with Zambia (100 MW), and some diesel generators will be retired (−38 MW);
- 2004: new gas turbine (75 MW);
- 2007: new hydrodevelopment at Rumakali (222 MW);
- 2009: new hydrodevelopment at Ruhudji (246 MW).

With the country's hydropower potential estimated to be in excess of 4700 MW (SAD–ELEC and MEPC, 1996, p 196), even more hydro projects could be part of the future. Coal and various renewables (particularly solar, wind and geothermal) could also generate more electricity in the future. For now, however, we include only those specifically mentioned by Tanesco in the baseline until and including the year 2010 for supply and 2025 for demand, which is outlined in Figures 4.2 and 4.3.

Zambia

Electricity production in Zambia is dominated by hydropower resources. Three large hydrodevelopments constituted over 90 per cent of the country's effective power capacity. These three are the Kafue Gorge station (with a capacity of 900 MW), the Kariba North station (on the Zambezi River, 600 MW) and the Victoria Falls station (108 MW). Four other smaller hydro stations are located in the northern system (servicing the mining area), which together have a capacity of almost 62 MW. Small gas turbine (80 MW), geothermal (20 MW) and diesel (4 MW) stations complete the generating picture (Eskom, 1996a, p 73 and ZESCO, 1994, p 37). Approximately 13 per cent of the total electricity produced during 1994/95 was exported (Eskom, 1996a, p 73).

Demand for electricity in Zambia is projected to grow only modestly during the coming two decades (primarily slowed by declining demand in the mining sector); for the entire 1997 to 2014 period, the average annual growth rate is just 1.8 per cent. Nevertheless, it is estimated that demand will surpass present supply sometime during the early part of the new century (ZESCO, 1997). To meet this, ZESCO could well develop the major remaining site on the Kafue River, bringing on 300 MW in 2003, followed by another 300 MW in 2005. These we include in our baseline, along with the demand projections over the coming 18 years (to 2014). Both are outlined in Figures 4.2 and 4.3.

Zimbabwe

The Zimbabwe electricity supply consists of a mix of coal-fired and hydropower stations. During 1994/95, the former produced 5526.3 GWh of electricity, and the latter 2284.7 GWh. Significant quantities had to be purchased from Zaïre and Zambia (over 1000 GWh in the case of each), while another 162.8 GWh were bought from the South African utility Eskom. In total, 9728.3 GWh were available for distribution in 1994/95 (Eskom, 1996a, p 75).

ZESA, the country's utility, expects demand to grow at between 3 and 5 per cent a year during the next two decades (in its baseline scenario). This would bring peak demand up to 4597 MW in 2017 (ZESA, 1997). To meet growing demand, the utility also expects a number of new projects to come on line (adapted from ZESA, 1997):

• 1997: an increase of 231 MW from Cahora Bassa in Mozambique;

- 1998: an increase of 84 MW from an upgrade of the Kariba hydropower facility;
- 2001: an additional 330 MW from the construction of Hwange 7, a coal-fired power station;
- 2003: another 300 MW from Hwange 8, another coal-fired power station;
- 2003: an additional 600 MW from the construction of a coal-fired power station at Sengwa;
- 2005: another 600 MW from Sengwa (a coal-fired power station);
- 2006: an increase of 150 MW from an extension to the existing hydropower facility at Kariba South;
- 2008: another increase of 150 MW from a second extension to Kariba South;
- 2010: an increase of 800 MW from the new Batoka hydropower facility on the Zambezi River.

This information allows us to complete demand-and-supply profiles for Zimbabwe, until and including the year 2017 for the former, and the year 2012 for the latter. These are outlined in Figures 4.2 and 4.3.

REVIEWING UTILITY DATA

We now have projections for supply and demand – largely gained from official sources – for the countries of the region. However, from a review of Figures 4.2 and 4.3, we immediately note that not all of our projections have the same time horizon; moreover, none of them go through to our final study year (2050). We call this a problem of missing information, and we return to it below.

In addition, however, to missing information, we are also concerned about inconsistent information. Consequently, we now turn to the information that has been provided in order to see if any of it appears problematic. Indeed, there are two elements that raise questions, and we examine each of them below.

The first is that there appear to be differences of opinion among utility planners, with respect to future electricity developments in the region. We come to this conclusion after observing that there are specific inconsistencies in the regional scenario that we have developed. For one, it is well known that Zimbabwe and Zambia have different expectations with respect to the Batoka hydropower facility: the former expects it to come on stream in 2010, while the latter has no definite plans. Given that the cooperation of both countries is vital to the efficient commissioning of the plant (both countries co-own the water rights), this inconsistency is worthy of attention.[8] Another inconsistency is that, in the year 2006, South Africa evidently expects to require all of its own domestic generation (in addition to another 1950

MW from Mozambique and Zambia) in order to satisfy its own demands. At the same time, however, Botswana expects to be importing 410 MW from South Africa. This may prove to be problematic.[9]

A second problem is that the countries of the region appear to be developing different visions, or at least different estimates, regarding future growth. Though these countries' economic (and electricity) prospects are not necessarily inextricably linked, it may be reasonable to expect their futures to be somewhat related. Indeed, given the common physical climate that many of the countries experience (and, related to this, the heavy dependence that many have upon agriculture), given the common global market in which many of their exporters compete, and given their growing economic integration, it might be useful to generate a common regional story which could form the basis for national electricity projections and planning. Consequently, the ongoing efforts to do so – in, for example, the SADC structures in general, and the Southern African Power Pool in particular (see the Appendix to Chapter 3) – are welcomed and encouraged.

Let us now return to this issue of incomplete information. The reader will recall (from Figures 4.2 and 4.3) that, beyond the year 2005, we are in the terrain of incomplete information. Indeed, by the year 2015, we have gone beyond the planning horizon of over half of the region's utilities. Nevertheless, our baseline requires demand-and-supply projections until, and including, the year 2050. We consider each in turn.

Developing the Demand Profile

Predicting electricity demand is, of course, awash with difficulties – indeed, this is the case for a decade or even a year into the future, let alone a half century. Nevertheless, in order to develop our baseline, we need to make distant projections. To do this, we make a key assumption and turn to a number of sources for guidance.

Our assumption is that, in most countries, growth in electricity demand will, at the outset of our study period (that is, the early part of the next century), be higher than growth in GDP. We assume this because history has often shown that, during the early phases of industrialization and urbanization, countries have witnessed rapid growth in electricity demand (with the movement away from traditional fuels to electricity in households being a key driver). At some point, however, a transition begins to occur, when growth in electricity will be less than growth in GDP (because of supply saturation, greater wealth bringing greater efficiency, technological developments and so on).

In order to attach specific numbers to these ideas, we turn to different sources. The African Development Bank, for one, suggests that energy consumption in southern Africa will rise by about 3.5 per cent annually during the coming decade (AfDB, 1996, p 47). Moreover, the World Energy

Council suggests that electricity use, in sub-Saharan Africa as a whole, could increase at an annual rate of 3.3 per cent between 1990 and 2020, and 3.2 per cent between 2020 and 2050. This is in the light of projected GDP increases (at market exchange rates) of 3.0 per cent and 3.5 per cent respectively (quoted in Lennon et al, 1996).[10] This fits with our assumption above.

Commenting upon these estimates from a South African perspective, Steve Lennon (after consulting within that country), concluded that:

> *Currently, there is a strong positive correlation between GDP growth and energy (in particular electricity) consumption. This trend is likely to continue for most African nations. It may become less apparent in South Africa as the energy intensity of the economy decreases with an increase in the benefication of raw materials. It is, however, recommended that the current electricity demand trend of GDP plus 1–2% be reflected in the study for Africa as a whole. The percentage of the energy share enjoyed by electricity is likely to increase as more Africans are given access to electricity. The rate of uptake of electricity is however strongly related to GDP growth rates.*
> (Quoted in Lennon et al, 1996)

For the purposes of our projections, we divide southern Africa into two areas. One is South Africa, which we assume is somewhat ahead of the rest of the region in terms of progress towards an 'energy transition'. Taking our cue from Eskom's own projections for its 'moderate scenario' (Eskom, 1996b), we suggest that the annual growth rate in peak demand in South Africa will be 2.0 per cent between 2015 and 2020, and 1.5 per cent thereafter (until 2050). The other area is the rest of southern Africa, which we propose will have an annual growth rate of electricity demand of 3.5 per cent until 2020, and 2.0 per cent between 2020 and 2050.

By using these growth rates to project peak-power demand values for periods beyond the utilities' planning horizons, we are able to extend the lines in Figure 4.2 out to 2050; the result is Figure 4.4. For the region as a whole, peak demand is 73.5 GW in 2030 and 101.7 GW in 2050. In each case, South Africa dominates this aggregate total, contributing 54.2 GW (73.7 per cent) and 73.1 GW (71.9 per cent) respectively. To give such values more meaning, we compare them with some present day systems. To meet this anticipated demand (including a reserve margin, which we discuss below), southern Africa would need, in 2010, a system that is about the size of present day Brazil's.[11] Additionally, southern Africa would need a system that is the size of present day India's in the year 2028, and that of present day Germany's in 2046 (Eskom, 1996a, p 84). To us, it seems reasonable that, in approximately ten, 30 and 50 years' time respectively, the size of the system in the entire southern Africa region could be thus.

Before finalizing these demand figures, we make one final amendment. While each country has a projected maximum demand (which we have reported in the previous section and have been developing since then), each

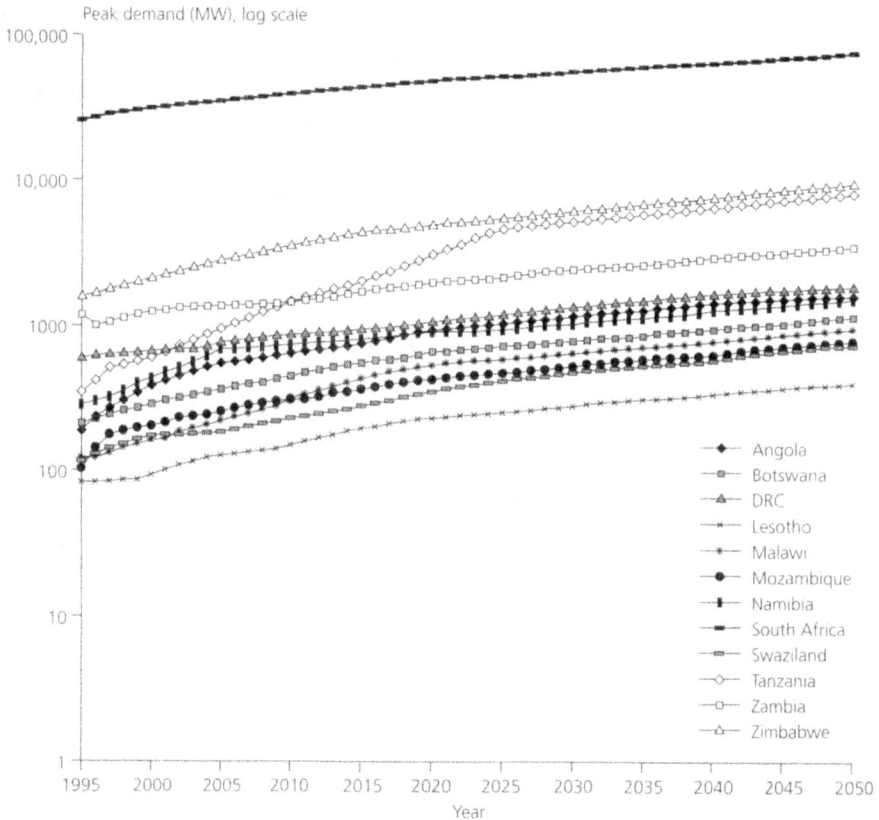

Figure 4.4 *Projected peak power demand in southern African countries,*
to the year 2050

country will also specify a reserve margin. This is the amount of generation
that is available to meet demand during unplanned and planned outages of
plant. The reserve margin is usually a function of, among other things, plant
load, generating characteristics and location of supply and demand. For the
sake of this study, the reserve margin will be assumed to be 20 per cent for
each country.[12] Therefore, to account for this, we inflate the demand projections by this amount.

Consequently, we now have demand, and inflated by our reserve margin of
20 per cent, we know what supply we have to fulfil. To create the baseline on
the side of supply options, we first identify all of the possible options in the
region for power generation. Let us consider them individually by resource.

Meeting Demand with Supply

Coal dominates regional electricity supply today in southern Africa, and the
quantity of reserves in the region suggests that it could do so for years to

come. According to the most recent issue of the *BP Statistical Review of World Energy*, South Africa's proved recoverable coal reserves amount to 55,333 million tonnes (British Petroleum, 1997), while *World Resources* reports that the country's proved reserves total 121,218 million tonnes (World Resources Institute, 1996, p 288). Other estimates suggest that there is enough coal in the country to sustain 100,000 MW of generating capacity (Lennon, 1996, p 8).[13] Additionally, other countries in the region with major proved coal reserves are Botswana (7000 million tonnes), Zimbabwe (1535 million tonnes), Swaziland (1000 million tonnes), the Democratic Republic of the Congo (720 million tonnes), Tanzania (304 million tonnes) and Mozambique (at least 240 million tonnes) (World Resources Institute, 1996, p 288).[14] Together, these resources could sustain at least 25,000 MW worth of coal-fired power stations for 35 to 50 years.[15]

There is also, however, considerable hydropower potential in the region. Table 4.1 lists the various potential projects in the region (in addition to those that have already been included in the utilities' plans, reported above). Although estimates of peak power potential vary considerably, additional

Table 4.1 *Hydropower potential in southern Africa (in addition, that is, to those resources already developed or already identified in utilities' plans)*

Location		Potential (MW)
Zambezi River Basin		
Kariba North Extension	Zambia	300
Batoka Gorge	Zambian side only	800
Devil's Gorge	Zambia/Zimbabwe	1240–1600
Mupata Gorge	Zambia/Zimbabwe	1000–1200
Cahora Bassa North Bank Extension	Mozambique	550–1240
Mepanda Uncua	Mozambique	1600–1700
Total Zambezi		approx 6000 MW
Other sources (non-Inga)		
Angola	including Kunene River Basin	16,400[16]
Lesotho		160
Malawi		250
Mozambique	other than the Zambezi	7250[17]
Namibia	other than the Kunene River Basin	500
Swaziland		60
Tanzania		3,000
Zambia	other than the Zambezi	1084–1308[18]
Total other sources (non-Inga)		approx 28,800
Inga		36,000–100,000
Total hydropower potential southern Africa		70,800–134,800

Sources: Black and Veatch International, 1996b; Dale, 1995; Dutkiewicz, 1996; Lwiindi, nd; SAD-ELEC and MEPC, 1996; and World Resources Institute, 1996

hydropower resources (excluding the Inga site) could add over 34,000 MW to peak supply in the region (though somewhat lower energy values, relatively, owing to the lower availability of hydropower). Full development of the Inga site in the Democratic Republic of the Congo, meanwhile, could add an additional 36,000 MW to 100,000 MW.

Southern Africa is also home to substantial gas reserves, both natural gas and coal-bed methane. Let us first consider natural gas. In terms of all potential resources (that is, not only producing areas and discovered areas, but also those areas that are referred to as prospective and speculative), Angola contains the most, with a total of 4284 billion cubic metres. Namibia is next with 1474 billion cubic metres, followed by Mozambique (298 billion cubic metres), South Africa (289 billion cubic metres) and Tanzania (97 billion cubic metres) (SADC, 1995a, p 2.8). Were these resources to be used for electricity production, a study from 1995 suggests that it would be possible to support 4750 MW of capacity in the early part of the 21st century, rising to 8950 MW by 2020 (SADC, 1995b, pp 5.5 and 5.21).[19]

Turning to coal-bed methane, it has been estimated that the region's extensive coal deposits contain between 900 and 3000 billion cubic metres (or 30 and 100 trillion cubic feet) of coal-bed methane (SADC, 1995a, p 2.10). From this, a back-of-the-envelope calculation suggests that, if we assume 50 per cent recovery (following SADC, 1995a, p 2.11), and that 0.5 billion cubic metres a year could sustain a 330 to 400 MW combined-cycle power plant (following Advanced Resources International, Inc. and Resource Exploration and Development, 1995), a peak power supply of 2970 MW to 11,880 MW could be sustained by southern Africa's resources over a 100-year period.

Nuclear power, moreover, must also be part of any comprehensive list of options. Given that South Africa has the capability to produce nuclear power – along with vast reserves of uranium (it was estimated, in 1987, that 144,400 tonnes were recoverable at less than US$80 per kilogramme; World Resources Institute, 1996, p 288) – we suggest that considerable electricity generation from nuclear power is a possibility in the future. Indeed, one estimate is that nuclear power could sustain power production totalling 53,000 MW (Lennon, 1996).

Finally, let us consider renewables. This is probably the most difficult resource to forecast, because the pace at which renewables will be developed is unclear. What is beyond question, however, is that the potential for renewable energy in the region (in addition, that is, to large-scale hydropower) is significant: solar, biomass and mini-hydro all present great possibilities.[20] The incidence of solar radiation in southern Africa, for example, is amongst the highest in the world; it averages 4.5kWh per square metre ('Coal-Fired South Africa', 1996).[21] Indeed, given its potential, all of the region's electricity needs could, in theory, be met by solar energy alone. The extent to which we should reasonably expect this to occur will be considered below.

Table 4.2 summarizes this examination. To this point, we have simply identified the resources available, but we have said little about the relative attractiveness of each, on whatever measure. In order, however, to construct a reasonable baseline, we must do just that – that is, we must anticipate upon what criteria different countries' decision-makers will select their resource type for meeting electricity demand. We begin this task by considering how others have foreseen the future.

Table 4.2 *Summary of southern Africa's electricity supply options*

Resource	South Africa	Rest of southern Africa
coal	100,000 MW	25,000 MW
hydro	small	34,800 MW (non-Inga)
		50,000 MW (Inga)
gas	350 MW (natural gas)	8600 MW (natural gas)
	6000 MW (CBM)	1500 MW (CBM)
nuclear	53,000 MW	small
renewables	significant	significant

Many argue that, although there might be some limited potential for growth in the use of natural gas for electricity generation, it is generally expected that coal and hydro will predominate in meeting southern Africa's future electricity needs. Justification for this is usually provided by references to costs (which we will consider in greater depth in the next chapter). For now, we simply state that, in a general sense, gas, nuclear and renewables are too expensive (comparatively) for electricity generation in southern Africa.[22] We accept this, but with one proviso.

In a recent scenario, the World Energy Council suggested that almost 30 per cent of sub-Saharan Africa's electricity needs could be met by renewable energy in the year 2050 (renewable energy in addition, that is, to hydropower) (quoted in Lennon et al, 1996). Given this and other similar observations, we would obviously be complacent if we did not justify our relative inattention to renewables.

As mentioned above, the extent to which renewables will be part of any energy future is fraught with uncertainty. While a case for virtually any figure could be made, we will assume that renewables play a modest role in southern Africa's electricity future: rising from a supply equal to 1 per cent of projected demand (including reserve margin) in 2030 to 5 per cent in 2050.[23] Based on these assumptions and assertions, the key question that remains is: 'What proportion of demand will be met by coal and what by hydro?' To try to determine a reasonable projection, we make four additional assumptions. Firstly, since hydropower is generally cheaper than coal in the region (again, this assertion will be further elaborated upon and defended in the next chapter), countries will, in the absence of other factors, choose to develop hydropower (in terms of building facilities on their territory, or paying neighbours to have facilities built on their territory, or building facilities themselves on their neighbours' territory). There will, however, be other factors; these are our three other assumptions.

We assume that South Africa will ensure that no more than 15 per cent of its projected demand (including reserve margin) will be met by imports. Moreover, when the import figure is near this maximum level, South Africa will want to secure a diversity of supply. Security of supply has been a part of the energy scene in South Africa for many years; the reasons for this during the apartheid era are probably self-evident.[24] It has, however, continued to be a key motif since the emergence of a democratically elected government in that country. Given persistent uncertainty in the region (in terms not only of politics, but also economics and technical factors), South Africa has been loathe to place the crucial issue of its electricity supply in the hands of others. For example, the country's National Electricity Regulator customer and support services general manager, Johan du Plessis, recently claimed that South Africa will not become too dependent on resources outside of the country for strategic reasons (Lourens, 1997).

Therefore, since the 'primary source of electricity in South Africa is likely to remain coal given the significant low cost reserves available' (Lennon, 1996, p 7), we assume that this means that the vast majority (projected demand, including reserve margin, less 15 per cent) of South Africa's future demand will be supplied by coal-fired power stations located within South Africa (less the nuclear and hydro capacity already operating in 1997).

Our third assumption is that the other large countries in the region – Tanzania, Zambia and Zimbabwe – will continue to exhibit similar desires for security of supply.[25] We also assume, however, that these countries may not be able to press their demands to the same extent (there may, for example, be pressure from donors to pursue cheaper imports). Consequently, we project that each of these countries will try to ensure that no more than 25 per cent of their projected demand (including reserves) will be met by imports.

Finally, our last assumption is technical. We assume that, all else being equal, there will be a desire to have supplies as close to loads as possible. Failing that, sources of supply should be fairly well distributed. Such arrangements help to promote technical stability within the system, and thus its reliability as well. We will also assume that excessive reliance on either thermal or hydropower, individually, is undesirable – given the different load nature of each (thermal, particularly when in terms of coal, often supplies base load, while hydro often meets peak demand).

Based on these assumptions, we create a supply profile; the year 2050 is presented in Figure 4.5. Let us briefly highlight the key features for each country.

- Angola: the country exploits its hydro resources by developing hydropower stations in 2015 (1500 MW) and 2042 (1500 MW).

Most of the electricity generated by these facilities is exported (with some, however, meeting rising domestic demand).

- Botswana: the country relies on imports to meet its rising demand; 1000 MW from 2016 to 2041, and 1300 MW after 2042.
- Democratic Republic of the Congo (DRC): the DRC begins to fulfil its potential as 'southern Africa's powerhouse', as hydropower facilities of 3000 MW each (primarily for export) are developed in the years 2010, 2033, 2039 and 2048. Total hydropower capacity at the end of our study period surpasses 14.5 GW.
- Lesotho: we assume that the country relies primarily on imports, rising to 200 MW in 2023, and to 300 MW in 2039.
- Malawi: the country develops its own hydropower resources further, by constructing a 100 MW facility in 2011 and a 150 MW one in 2014. It then pursues imports from 2020 (200 MW in that year, rising to 600 MW by 2050).
- Mozambique: it develops Mepanda Uncua (1700 MW) in 2010, and then Cahora Bassa North Bank Extension (1240 MW) in 2014, mainly for export. It also develops a large 3000 MW facility in 2025, again primarily for export.
- Namibia: imports increasingly dominate the supply profile in Namibia, rising every decade (to 500 MW in 2024, 800 MW in 2033 and 1000 MW in 2045).
- South Africa: rising demand is met by a combination of 3000-MW coal-fired power stations (with a new one coming on line every three to six years) and, to a lesser extent, imports, with an additional 2000-MW import agreement concluded approximately every decade.
- Swaziland: additional supply is provided exclusively by imports, rising in 200-MW gradients, to reach 900 MW by the end of the study period (2050).
- Tanzania: the country exploits its own resources first, building new hydropower plants, each with a capacity of 1000 MW in 2011, 2015 and 2020. After taking on some imports (500 MW in 2023, followed by an additional 500 MW in 2025), the country turns to its coal resources and constructs a 1000-MW thermal plant in 2028. Beyond this, however, it must rely on imports – an additional 500 MW in 2038, another 1000 MW in 2042 and a final 500 MW in 2050. (At the end of the study period, the country has surpassed its desired limit on imports, but this was necessitated by the lack of any further domestic resources for electricity generation.)
- Zambia: the country develops its hydroresources (for export, and then increasingly for its own needs) – initially those that are wholly within its own territory (for these are cheaper and easier to manage), and then those shared with Zimbabwe. Specifically, this is 1000 MW (most probably in the Luapala River Basin) in 2021. Then, in conjunction with Zimbabwe's activities, it develops 800 MW at Devil's Gorge (in

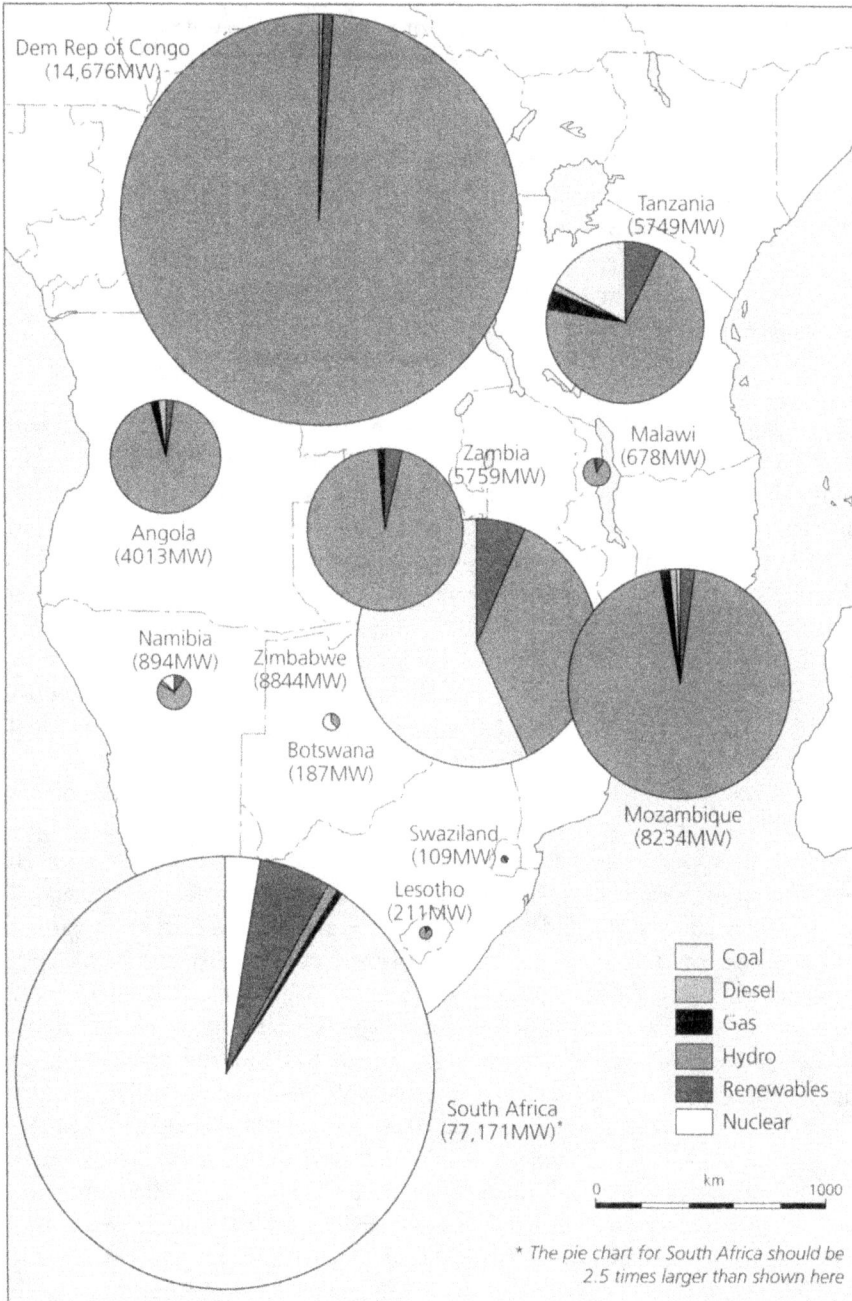

Dem Rep of Congo
(14,676MW)

Tanzania
(5749MW)

Malawi
(678MW)

Zambia
(5759MW)

Angola
(4013MW)

Namibia
(894MW)

Zimbabwe
(8844MW)

Botswana
(187MW)

Mozambique
(8234MW)

Swaziland
(109MW)

Lesotho
(211MW)

South Africa
(77,171MW)*

Coal
Diesel
Gas
Hydro
Renewables
Nuclear

km
0   1000

* The pie chart for South Africa should be
2.5 times larger than shown here

Note: The fact that renewables contribute a different percentage share to national generating capacity in different countries does not suggest that some countries will be more adept at developing renewables than others. Instead, it is simply a consequence of the fact that renewables satisfy the same percentage share of demand in each country. Therefore, countries that rely heavily on imports (such as Botswana and Swaziland) will have a (relatively) larger contribution of renewables to their (relatively) smaller total national generating capacity.

Figure 4.5 *Projected generating capacity in southern Africa,
by resource and country, 2050*

2027) and 600 MW at Mupata Gorge in 2043. The one exception to this is its willingness, we assume, to develop its share of Batoka (800 MW) in 2010 (at the same time, that is, as Zimbabwe develops its 800 MW share).

- Zimbabwe: the country meets its growing demand with a combination of imports, amounting to 1000 MW in 2015, another 1000 MW in 2036 and a final 500 MW in 2046, thus always keeping close to the 25 per cent ceiling. The shared projects with Zambia are (as noted above) 800 MW at Devil's Gorge (in 2027) and 600 MW at Mupata Gorge in 2043. Finally, it constructs a 1000-MW thermal plant in 2019, followed by 500 MW facilities in each of 2033 and 2049.

Taken together, southern Africa's generating capacity in the year 2050 will be 60.3 per cent coal, 32.6 per cent hydro, 1.5 per cent nuclear, 4.8 per cent renewables, 0.6 per cent gas and 0.2 per cent diesel. Compare this to the situation in 1995, when the system was 78.0 per cent coal, 15.9 per cent hydro, 4.2 per cent nuclear, 1.4 per cent gas, 0.5 per cent diesel and no renewables. Therefore, although there has been a clear shift away from coal and towards hydro (hydro has more than doubled its share, and increased its absolute contribution more than sixfold), the system is still dominated by coal-fired power (over three-fifths of all generating capacity). The accompanying increase in greenhouse gas emissions is considerable, but we will consider this in further depth in the next chapter.

It is also worth noting that regional cooperation is already an important part of our baseline: indeed, in the year 2050, 20,550 MW of capacity are used to meet electricity needs in another country. In other words, over 16 per cent of all electricity generated is destined for international markets. It is, of course, only right and proper to have regional cooperation as part of our baseline: the recent progress made in the Southern African Power Pool demands this (see the Appendix to Chapter 3). We highlight it here simply to remind the reader that, although our mitigation scenarios exploit regional opportunities, we are not suggesting that our baseline scenarios are built upon the desire for complete national self-sufficiency in electricity supply.

To complete this chapter, we acknowledge that not everyone will necessarily accept our baseline. Indeed, it is founded on a number of assumptions, a small change in which could shift the picture in the year 2050 dramatically.[26] Nevertheless, we hope that it is defensible, for the various reasons we have explained. Moreover, we believe that the methodology is sufficiently transparent to allow those dissatisfied with the baseline the opportunity to develop their own. For now, however, we adopt this as our baseline and investigate possible mitigation options in the next chapter.

ENDNOTES

1 The two SADC members we do not examine are Mauritius and the Seychelles. We examine the 12 mainland countries of SADC because of their geographical proximity to each other, because of the complementarities that exist among their electricity supply-and-demand profiles and because of the institutional steps they have already taken in electricity cooperation (see the Appendix to Chapter 3). These points will be expanded in this chapter, as well as in the next two chapters.

2 Our analysis in this book focuses upon peak loads (power). We nevertheless accept that a comprehensive analysis of future electricity demand and supply must consider both peak loads and total electricity generated. (This is because, for one, a 1000-MW hydrofacility may be able to generate much less electricity over the course of a year than a 1000-MW coal-fired power station. This would be in spite of the fact that each might be able to satisfy the same peak demand, at least for a short period of time.) Given the exploratory nature of our study, however, we believe that we are justified in concentrating only on peak loads. We return to this point in the next chapter.

3 We recognize that the peak loads in the different systems did not necessarily occur simultaneously.

4 This study, by the World Bank Industry and Energy Operations Division, is entitled *People's Republic of Angola, Power Sector Rehabilitation Project*. The two non-governmental organizations are Southern African Development Through Electricity (SAD–ELEC) and Minerals and Energy Policy Centre (MEPC).

5 Obviously, such a project – were it to be implemented – might make use of power potential that we subsequently introduce into our baseline or our mitigation scenario. We could, therefore, run the risk of making use of a resource that had already been exploited. We will be attentive to such a possibility.

6 There is, however, much controversy about the potential impact that such a project would have.

7 'Pumped storage facilities are net users of electricity and are used for peaking. Water is pumped during off-peak periods to generate electricity during peak periods' (Eskom, 1996a, p 10). For the purposes of this study, we will consider the small amount of electricity available from 'pumped storage facilities' together with that available from coal-fired power stations.

8 This is not to say that Zimbabwe could not pursue this project virtually on its own. One consultant's report, however, concluded that each country should pursue the project simultaneously (each installing 800 MW of turbines on its own side) in order to maximize the resulting benefits (Batoka Joint Ventures Consultants, 1993).

9 Indeed, a number of countries in the southernmost part of the region expect to be importing from South Africa during the early part of the next century.

10 Unless otherwise noted, any figures used from the World Energy Council's study will refer to their 'B' scenario. This projects modest economic growth and technology development, and has been labelled 'most likely to portray the African situation' by many South African analysts (Lennon et al, 1996).

11 Or at least Brazil's, as it was in 1993.

12 This figure is not universally agreed upon. A consultant studying Tanzania's electricity future, for example, used a reserve requirement of 15 per cent (Black and Veatch International, 1996b, section E-7). Moreover, the Zimbabwean utility ZESA has a 'minimum reserve level ... (that is) at least 20 per cent of adjusted

demand (adjusted demand is equal to the peak system demand plus the amount of firm tariff power exports minus the amount of firm tariff power imported in the same interval)' (Black and Veatch International, 1996a, p 2.13). An Eskom official, studying future prospects for South Africa, notes that a 12 per cent reserve margin may be 'possibly optimistic' (Lennon, 1996, p 7). Eskom itself quotes a 15.2 per cent figure for 'excess above gross plant margin' (Eskom, 1996b, p 4). Finally, a consultant reported that the Southern African Power Pool was promoting '19 percent planning reserve margin for an all thermal utility, 10 percent for an all hydro utility, and a weighted average for a mixture of thermal and hydro' (Black and Veatch International, 1996a, p 2.13).

13   The time horizon for this estimate is not given.

14   'Tanzania is endowed with coal reserves estimated to be as high as 1,200 million tonnes, of which 300 million tonnes are proven' (Black and Veatch International, 1996b, section E-4).

15   One study found that the 159 million tonnes of coal (proved reserves) in the Mchuchuma and Katewaka coal fields could fuel a power plant of up to 400 MW capacity for 35 to 50 years (Black and Veatch International, 1996b, sections E-24 to E-25). We extrapolate from this to arrive at the 25,000 MW figure.

16   Relatively little detailed work has been done on Angola's considerable hydropower potential, mainly because of the civil war in the country. 'The shared Cunene river has been estimated to have the potential to generate in excess of 2,000 MW of hydroelectric power, although any development will have to be negotiated [by Namibia] with Angola' (SAD–ELEC and MEPC, 1996, p 172).

17   This includes the Alto Malena, the Lupata I and II and Boroma sites.

18   This includes the Luapala River Basin, the Luangwa River Basin (Lusiwasi Extension) and other sites on the Kafue River Basin (Itezhi–Tezhi).

19   Discussions about gas have moved beyond the theoretical, at least tentatively: the South African utility Eskom, and the developers of the Kudu field in Namibia, have held discussions about the possibility of constructing a 1750 MW power plant. As of early 1998, however, no agreements have been reached. Others, moreover, report that these fields could sustain a 4000 MW power station for more than 50 years (SAD-ELEC and MEPC, 1996, p 73).

20   One estimate claimed that wind power in the region had the potential to produce 1900 MW ('Coal-Fired South Africa', 1996).

21   Peaks are, of course, even higher. In Zimbabwe, between November and January, the levels (depending on the rain pattern) are 6.5kWh/m² (Southern Centre, 1997, p 2.4).

22   Supporting this, in the World Energy Council's scenario for sub-Saharan Africa as a whole, no natural gas was projected to be used in electricity generation (quoted in Lennon et al, 1996). Moreover, an Eskom official was quoted as saying that 'Nuclear power generation is a highly unlikely option due to this process being extremely costly' (quoted in Lourens, 1997).

23   Renewables are also examined in Chapter 7 of this book.

24   Security of supply is, of course, a well-known concept in energy studies. To ensure a continuous supply of different energy requirements, states are willing to pay a premium for domestic, rather than foreign, supplies.

25   In 1993, for example, it was observed that 'the declared policy of the Zimbabwean government is to limit the import of electricity to a maximum of 25% of the national electricity demand' (Batoka Joint Venture Consultants, 1993, p 6.22). Three years later, however, this limit was reported to be 20 per cent by another consultant (Black and Veatch International, 1996a, p 2.13).

26 Changing, for example, South Africa's future growth in demand shifts the final numbers considerably. Consider the following figures:

Annual growth rate in demand from 2020–2050	Demand in 2050
1.5 per cent (as assumed in this study)	73,055 MW
0.5 per cent	54,281 MW
2.5 per cent	98,036 MW

Chapter 5 | Regional Electricity Mitigation Options

*Norbert Nziramasanga, Bothwell Batidzirai
and Ian H Rowlands*

INTRODUCTION

In this chapter, we investigate the potential for climate change mitigation in southern Africa by means of regional cooperation in electricity generation. We examine how the countries of the region could organize their electricity supplies in ways that minimize emissions of carbon dioxide, while still maximizing the performance of the system as a whole. Moreover, we also calculate the potential costs of such an arrangement (compared with the baseline or business-as-usual scenario laid out in the previous chapter).

The chapter is divided into three main sections, each exploring possible future developments for the region's power sector. The first examines what we have called our 'hydrophilic scenario'. To create this, we envisage an arrangement where the region's hydropower potential is fully exploited as soon as is needed (and is reasonably possible). The sequencing of hydropower projects is laid out, as is the resulting regional power supply and demand during the coming half century. Data concerning carbon dioxide emissions and associated costs (compared with our baseline scenario) are also presented. Finally, we discuss the desirability – primarily from a technical perspective – of this hydrophilic scenario.

The second main section of this chapter presents what we have called our 'hybrid scenario'. To create this, we have made use of both the region's hydropower resources and its (admittedly more limited) gas resources in order to develop a scenario that not only advances climate goals, but also addresses some of the more technical concerns that we will see are raised by our hydrophilic scenario. It is our assertion that this hybrid scenario strikes a more constructive balance between climate concerns and performance requirements than does the hydrophilic scenario.

In the third section, we investigate a few representative mitigation projects, which are also part of our broader scenarios. We do this in recognition of the fact that future mitigation activities may well be undertaken on a project-by-project basis (while, nevertheless, still being part of a broader planning scenario of some kind). Therefore, we believe that it is important to initiate analysis of individual projects as well. The extent to which it might be possible to actually implement either of our two scenarios – or, perhaps more importantly, the parts thereof (the individual projects) – will be examined in the next chapter.

HYDROPHILIC SCENARIO

Southern Africa's vast hydropower resources appear to offer great potential for the mitigation of climate change: climate-friendly hydroelectricity could displace climate-changing coal-generated electricity. In this section, we take such a potential to its reasonable limits by developing what we call our hydrophilic scenario – that is, a scenario in which the region's decision-makers choose to use hydropower resources to the greatest extent that is reasonably possible.

To begin with, we recall our 'demand' baseline that we developed in the previous chapter (see Figure 4.4). To meet this demand, however, we relax the various assumptions that we also laid out in that chapter (which, in fact, effectively acted as constraints on resource selection). Consequently, we will no longer require particular countries to have restrictions upon the quantity of electricity that they can import. Moreover, we will explicitly pursue hydropower resources: when new generating capacity is required anywhere in the region, it will be supplied by hydropower resources in the region, irrespective of the geographical relationship between supply and demand (though, all else being equal, we will try to keep them as close as possible, and to vary the sources). Notwithstanding these constraints, we will try to ensure that this scenario could, conceivably, come to fruition during the next half century.

Hydropower, of course, dominates the resultant hydrophilic scenario:[1] new demand for electricity is supplied, until the year 2048, exclusively by hydropower.[2] During this period, over 65,000 MW of new hydropower capacity comes on line (compared with 34,000 MW in our baseline scenario). Most of this is accounted for by the large-scale projects developed in the Democratic Republic of the Congo, as well as in Angola. Though other projects are also brought forward, the 39,000 MW developed in these countries (compared with 10,500 MW in the baseline scenario) constitute all of the new capacity. Table 5.1 summarizes the key differences between the two scenarios in hydropower development. Figure 5.1, meanwhile, shows the quantity of both coal-fired and hydro-generating capacity in this scenario, and for the baseline scenario as well. Generating capacity for the other resources – gas, diesel, nuclear and renewables – is approximately the same in each of the two scenarios.

We are confident that this hydrophilic scenario could actually be realized. Correspondence from Electricité de France, who were involved in a feasibility study of the Inga site, outlined a scenario in which approximately 6000 MW

Table 5.1 *Differences in hydropower developments, between baseline and hydrophilic scenario*

Location	Size (MW)	Year of commissioning Baseline scenario	Hydrophilic scenario
Angola	1,500	2042	
	3,000		2019
	3,000		2036
	3,000		2042
DRC	3,000	2033	
	3,000	2039	
	3,000	2048	
	6,000		2021
	6,000		2025
	6,000		2032
	6,000		2038
	6,000		2043
Mozambique	1,700	2010	2007
	1,240	2014	2013
	3,000	2025	2015
Zambia	1,000	2021	2007
	800	2027	2019
	600	2043	2025
Zimbabwe	800	2027	2019
	600	2043	2025

were brought on line in the DRC in each of years 2010, 2017, 2023 and 2030 (EdF, 1997). We are being more ambitious than this in only some respects – more specifically, with regard to total capacity and the rate of growth of generating capacity. Moreover, we are much less ambitious than are the plans outlined by consultants for the African Development Bank. In their 'regional co-operation scenario (with Inga)', they envisaged the development of four 3000 MW tranches in year 2005 and four more 3000 MW tranches in 2010 (AfDB, 1996, p 49). What we have not considered, however, is how Egypt might compete with southern Africa for the DRC's power supply (compare with, for example, AfDB, 1996). For now, we assume that electricity generated in the DRC can be wholly utilized in southern Africa.

Therefore, although some individuals have estimated southern Africa's hydropower potential to be well over 100,000 MW (see Table 4.1), we have restricted ourselves to 72,700 MW in total. Indeed, we reach this upper limit in year 2043 of our scenario, and, as a consequence, the region sees the construction of a new coal-fired power station in 2048 in South Africa, the first such event for 40 years. Until that time, South Africa had met its rising demand by relying upon hydropower capacity in other countries – to such an extent that imports satisfy over 50 per cent of its projected demand (including reserve margin) in years 2044 and 2045.

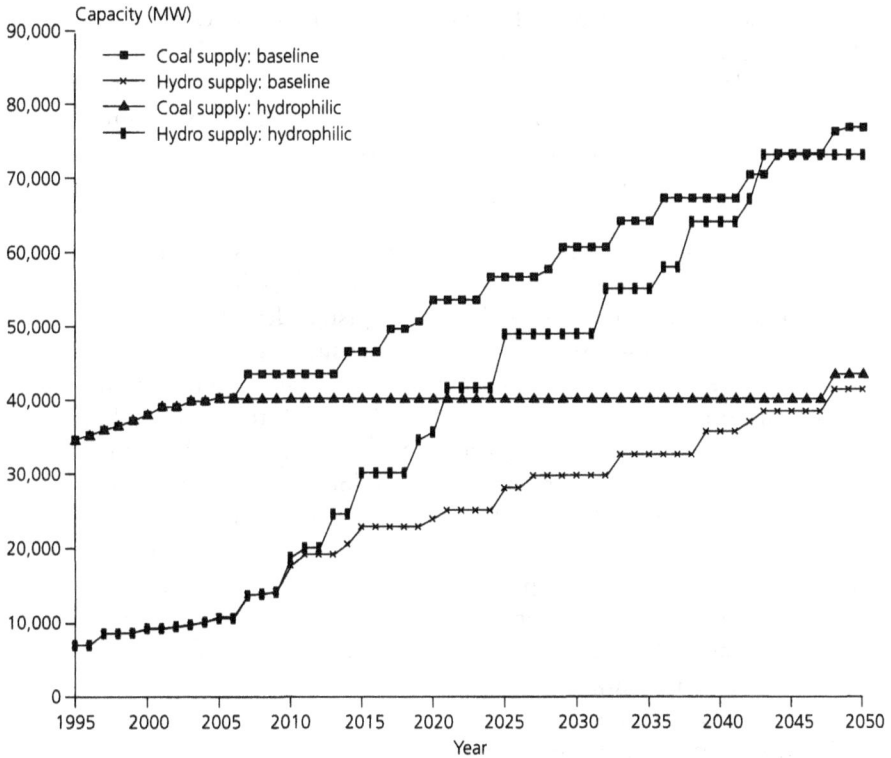

Figure 5.1 *Coal-fired and hydropower capacity in southern Africa, hydrophilic scenario and baseline*

In this hydrophilic scenario, the regional system is becoming increasingly hydrobased: in 1995, the respective shares were 78.0 per cent coal-based and 15.9 per cent hydro; in 2046, the shares become 33.8 per cent coal and 61.0 per cent hydro. Because of the need to return to coal as the main resource after reaching the upper limit of the region's hydropower potential, the shares become, in 2050, 34.6 per cent coal and 58.1 per cent hydro (with, furthermore, 1.5 per cent nuclear, 4.9 per cent renewables, 0.6 per cent gas and 0.2 per cent diesel). Nevertheless, hydropower is still the major supplier to the system (see Figure 5.1).

To calculate the quantity of carbon dioxide abated, assume first that hydropower produces no greenhouse gases. Though this is controversial – the construction of dams involves the emission of greenhouse gases, and many argue that dam reservoirs themselves may be significant producers of greenhouse gases (see, for example, Rosenberg et al, 1997, pp 32–34; and Gagnon and van de Vate, 1997) – we nevertheless still maintain that carbon dioxide emissions will be relatively modest. Assume, secondly, that all those power stations that would have been constructed in the baseline scenario (the difference between this mitigation scenario and our baseline) would have operated at 35 per cent efficiency, used coal with an emission factor of 94.6 kilogrammes of carbon dioxide per gigajoule (UNEP, 1997, p 102) and had a load factor of 70 per cent.[3] In all, there are 14 coal-fired power stations which, envisaged in our baseline scenario, are not built in our hydrophilic scenario: ten are in South Africa (3000 MW each), three in Zimbabwe (one at 1000 MW and two at 500 MW each) and one in Tanzania (1000 MW).

The quantity of carbon dioxide not emitted is considerable. In the year 2007 – the first year for which there are differences in emissions between the two scenarios – the savings amount to just under 18 million tonnes of carbon dioxide. These rise steadily during the subsequent years, reaching 77 million tonnes in 2020, 119 million tonnes in 2030, 158 million tonnes in 2040, and finally peaking at just under 197 million tonnes in each of years 2049 and 2050. Savings of this magnitude are equal to the total energy-related carbon dioxide emissions in all of Indonesia in the year 1992. Given that Indonesia ranked 23rd in the world, in terms of absolute emissions in that year, the quantity of greenhouse gases involved is significant (World Resources Institute, 1996, pp 326–327). Total savings for the entire study period amount to 4755 million tonnes of carbon dioxide. If we even consider this on an annual basis (over 44 years), savings are of the order of 108 million tonnes annually. This is still notable, for it is equivalent to over 10 per cent of Germany's energy-related carbon dioxide emissions in 1992 (World Resources Institute, 1996, p 326). Figure 5.4, located further on in this chapter, presents this information graphically.

However, at what price do we achieve this? Let us move on to consider the costs of such a scenario. To do this, we examine the difference in costs between the baseline scenario (developed in the previous chapter) and the hydrophilic scenario (developed above).[4] We focus upon those two resources that are used to different extents – namely, coal and hydro. First we consider the cost of the coal-fired power stations not constructed. As noted above, there are 14 of them. Additionally, one other coal-fired power station is constructed two years earlier in our hydrophilic scenario (a South African one in 2048 instead of 2050, as in the baseline). Data about the cost of new coal-fired power station plant in South Africa are not readily available.[5] We do, however, have some information about other proposed coal-fired power stations in the region, as well as estimates that others have made. They are listed in Table 5.2 below.

From this, we assume that the cost of new coal-fired power station facilities in South Africa will be US$1000 per installed kW (including transmission lines).[6] We support this by noting that the units would be relatively large (thus exploiting economies of scale), and that the South Africans, having had much experience with coal-fired generation, would have been able to perfect – or at least advance considerably – the techniques they use. Meanwhile, for the other countries developing coal-fired power stations in our scenarios, that is, Tanzania and Zimbabwe, we assume that new plant

Table 5.2 *Cost of new coal-fired power stations in southern Africa, various sources*

Site	Size	Cost per installed kW (US $)	Source
Sengwa, Zimbabwe	four 200 MW turbines	1,344	Black and Veatch International, 1996a, p 229
Sengwa, Zimbabwe	two 675 MW turbines	993	Black and Veatch International, 1996a, p 229
Mchuchuma/Katewaka, Tanzania	100 MW	1,237	Black and Veatch International, 1996b
'new thermal plant in South Africa'		1,000	AfDB, 1996, p 54
'new coal-fired plant with FGD'		1,300	World Bank IENOG, 1996

will cost US$1500 per installed kW because we believe that future plant should be priced somewhat higher than the anticipated Sengwa facility in Zimbabwe.

The other major costs for a coal-fired power plant will be, first, the fuel and, second, the other annual operating costs (such as staffing and maintenance). For fuel, we estimate the cost of coal to be US$0.67 per gigajoule (GJ) in 1997, falling to US$0.63 per GJ in 2050.[7] Given that we are using coal with a carbon dioxide emission factor of 94.6 kilogrammes per GJ, this is equivalent to almost US$26 per tonne in 1997, falling to US$24.40 per tonne in 2050. To calculate the amount of coal required, we make the same additional assumptions — that is, 35 per cent efficiency and a load factor of 70 per cent. For the other annual operating costs, meanwhile, we assume that they will be equal to, annually, US$15 per installed kW.[8]

Turning to hydropower, recall first the changes in the profile of hydropower development in the region (see Table 5.1 above). There have been numerous estimates for the construction costs of new hydropower facilities, many of them undertaken in Zimbabwe, Mozambique and Zambia. Some are listed in Table 5.3 below.

Based on Table 5.3, with regard to those facilities for which we have estimates — namely, the two projects shared between Zimbabwe and Zambia (Devil's Gorge and Mupata Gorge), the Luapala site in Zambia and the Mepanda Uncua site in Mozambique — we use the cost figures cited. For those facilities which do not have estimates, we will make our own.

For the additional 3000 MW site in Mozambique that we envisage as part of the hydrophilic scenario, we estimate the cost to be US$1250 per installed kW (including the necessary transmission lines). In the case of Angola, where relatively little research on potential hydropower development has been

Table 5.3 *Cost of new hydropower stations in southern Africa, various sources*

Site	Size	Cost per installed kW (US $)	Source
Batoka Gorge, Zambezi	1,600 MW	716	Batoka Joint Venture Consultants, 1993, p 14.13
Batoka Gorge, Zambezi	800 MW	1,375	Black and Veatch International, 1996a, p 229
Devil's Gorge, Zambezi	1,000 MW	787	Batoka Joint Venture Consultants, 1993, p 14.13
Mepanda Uncua Hydro Power Station, Zambezi	2,500 MW	800 (including transmission lines)	Eskom, 1998
Mupata Gorge, Zambezi	1,085 MW	985	Batoka Joint Venture Consultants, 1993, p 14.13
Luapala, Zambia	884 MW	713	Batoka Joint Venture Consultants, 1993, p 14.10
Kafue Gorge, Zambia		556	Eskom, 1998
Kafue Gorge, Zambia		696	Batoka Joint Venture Consultants, 1993, p 14.10
'Hydropower in Zambia and Zimbabwe'		1,000	AfDB, 1996, p 41

completed, we assume the costs to be US$1500 per installed kW, which is slightly higher than estimates for neighbouring Namibia (US$1200 per kW by AfDB, 1996, p 41; note as well that Eskom reports the cost of the Capanda Hydro Power Station to be US$1154 per kW; Eskom, 1998). For all of these figures, we assume that the costs of transmission lines are included.

Turning to the largest component of our hydrophilic scenario, a feasibility study for the Inga site concluded that construction of generating capacity would be 'less than US$500 per installed kW' (EdF, 1997). We will use a figure of US$500 per kW in our investigation. With regard to transmission, we assume that, for each 3000 MW facility we envisage constructed, new transmission lines will be required. Following estimates from studies by both Electricité de France (EdF, 1997) and the African Development Bank (AfDB, 1996), we assume that cost of transmission lines to South Africa would be equivalent to US$350 per installed kW, thus increasing the capital costs by 70 per cent. Finally, for all hydropower projects, we will assume that annual variable costs will be equal to US$10 per installed kW.[9]

To calculate the net present value (remembering that all cost estimates are in 1997 dollars), we use a discount rate of 5 per cent.[10] After calculating the difference in the costs of the baseline scenario and the costs of the hydrophilic scenario, we find that the net present value of these differences is 'negative' US$8.1 billion – that is, the hydrophilic scenario is cheaper than the baseline scenario. This should not come as a surprise: the annual costs of hydropower operation have been assumed to be lower than coal, and so too have the costs of new facilities (at least those in the Democratic Republic of the Congo, which dominate this scenario). Therefore, we are actually reducing carbon dioxide emissions at a net saving: US$1.71 per tonne of carbon dioxide. We maintain, however, that our consideration of costs should not end at this point.

We need to consider some of the technical issues that arise from the fact that this scenario relies heavily upon hydropower. Most importantly, because the region's electricity supply system will be highly dependent upon the region's hydrology, we need to investigate the reliability of the same. Drought is, of course, always a real possibility in southern Africa (and, moreover, its incidence could well increase given global climate change). Indeed, during the early 1990s, the Kariba Dam on the Zambezi River was reduced to less than half its maximum capacity, with a constant threat of losing supply completely if the rainy season was delayed by more than six months in the latter parts of 1992 and 1995. Thus, the chance that the hydropower facilities will not be able to provide the promised electricity is real and should always be part of planning.[11]

The cost of system reliability can be expressed in terms of the monetary cost of loss of supply, both planned and emergency outages.[12] Let us consider a situation where the hydropower facilities are only able to deliver half of their promised peak power (and thus electricity). If this were adequately covered by the reserve margin, then we would assume that the other existing plant could work to a higher load factor and ensure that there are no power shortages.[13] If, however, the outages are greater than the reserve margin, the system will not be able to satisfy all of the demand, even when working at 100 per cent load factors. Consequently, losses will occur because of two phenomena: first, because the system's capacity will be reduced by a specific quantity (amount of lost hydropower minus reserve margin), which we assume will be predictable and thus planned; and second, because the system that remains will experience unplanned outages, which will demand emergency load-shedding.[14] We consider each of these in turn.

What we call a 'planned outage' arises when the utility has given prior notice of outages. Given, for example, a low level of water in each of the dams' reservoirs – along with the fact that no more rain is expected for a time – a utility's representatives would fully realize that they will only be able to provide half the energy that they had previously expected. Anticipating this, they would implement restrictions, around which their customers would plan their operations. We assume that the value of lost energy in such an example would be equivalent to US$0.82 per kWh. We arrive at this figure by dividing the region's 1994 GDP (US$165 billion) by its energy consumption (201 terawatt-hours (TWh)). This, we believe, is a reasonable estimate of the value of planned outages.

It is, however, the unplanned outages that are much more costly. In this instance, we assume a situation in which a plant breaks down and 'blackouts' result. The value of such lost energy is much higher, since the utilities' customers are unprepared for it. We estimate the cost of such unplanned outages to be US$50 per kWh. We also assume that the region's power facilities are in good condition, so our loss of load probability is only five hours per year.[15]

Figure 5.2 presents the value (in 1997 dollars) of the outages during the study period. Note, first of all, that under the conditions described, there would be no outages in our baseline scenario (half of the hydropower generated is always less than the difference between total supply and total demand in the region). Turning to our hydrophilic scenario, however, we find that outages do arise in years 2018, 2020 and from 2023 onwards (at this point,

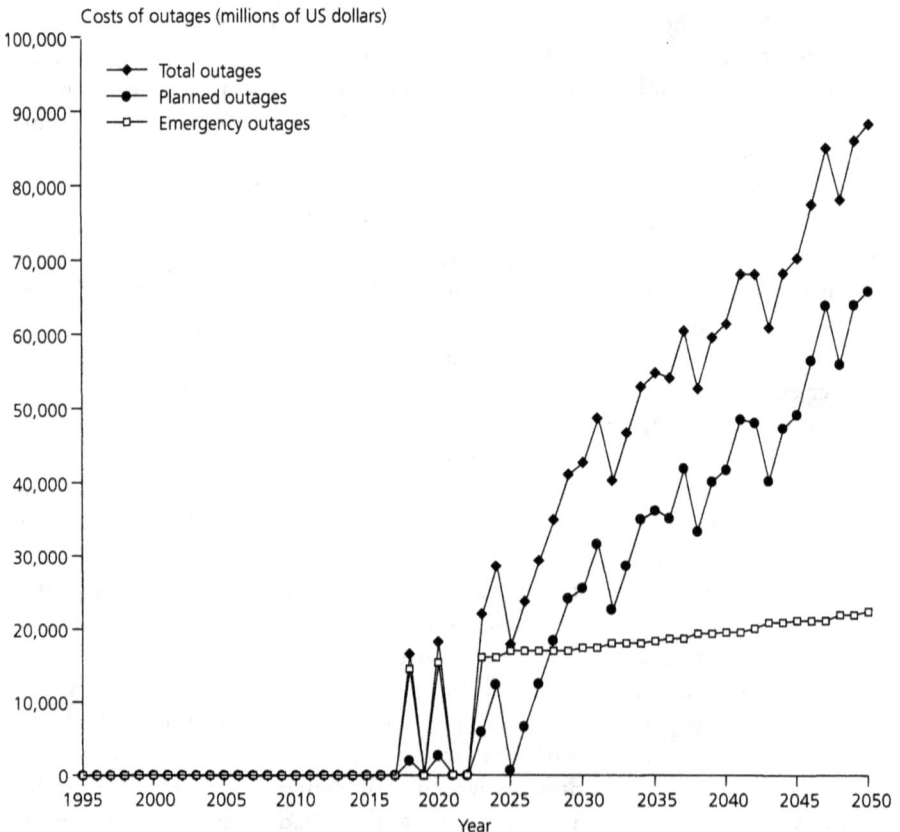

Figure 5.2 Cost of planned and unplanned outages arising from drought, hydrophilic scenario

lost hydropower is greater than the system's reserve margin). Outages rise steadily, to a value of US$88 billion in 2050 (about three-quarters attributable to planned outages and one quarter to emergency outages).

Because we believe it is crucial to add this to the cost of our hydrophilic scenario, we calculate the net present value of the costs of these outages and then add them to our already calculated capital and variable costs. We see that they increase the costs of this hydrophilic scenario (when compared with the baseline scenario) considerably – indeed, to such an extent that what was originally a net saving has now become a net expense. The cost of abatement is US$41.80 per tonne of carbon dioxide (with a 5 per cent discount rate). This compares with a pre-outage saving of US$1.71 per tonne of carbon dioxide. We take this figure to be our estimate for the cost of this hydrophilic scenario.

HYBRID SCENARIO

Dramatic reliance upon hydropower potential – as exemplified by our hydrophilic scenario, outlined above – brings with it technical difficulties which we have shown can be quite costly. In this section, therefore, we aim to develop a scenario which still makes use of the region's hydropower potential for climate gains, but does so in a more restricted manner. Moreover, it exploits the hitherto unused gas resources in the region. We call it a hybrid scenario, to reflect the different resources in the region's supply mix.

To generate our hybrid scenario, we impose a number of conditions on the development of new generating supply:

- The contribution of hydropower to the southern African system, as a whole, will be no greater than 40 per cent. This condition is introduced in order to ameliorate the problems, outlined above, of having a system made up predominantly of hydro. At the same time, however, we still want to capture the climate advantages and some of the economic benefits of this mode of electricity generation. Introducing a constraint of this kind might, we believe, address both issues.
- The contribution of hydropower imports to South African demand (including reserve margins) will be no more than 20 per cent. In addition to our first condition, this is intended to reflect the particular concern that South Africa has about availability of supply. Moreover, it is a concern which South Africa could address: namely, turn to domestic coal for power generation. This is, we believe, a reasonable constraint which falls between the 15 per cent limit (on total imports) in our baseline scenario, and the lack of a limit in our hydrophilic scenario. For their part, Tanzania and Zimbabwe still aim for a 25 per cent ceiling on total imports.

- The use of gas (both natural gas and coal-bed methane) is preferred to coal for electricity generation.

The hybrid scenario, full details of which can be found in UNEP (forthcoming), makes use of a wider range of resources. Figure 5.3 shows the quantity of coal-fired, gas-fired and hydrogenerating capacity in this scenario, and for the baseline scenario as well. Generating capacity for the other resources – that is, diesel, nuclear and renewables – is approximately the same in each of the two scenarios. The 50-plus-year study period can be generally divided into three separate phases. The first runs until 2014. During this period, there are few constraints upon the system, and hydropower developments proceed more quickly than in the baseline scenario – the Cahora North Bank Extension in Mozambique, for example, comes on line in 2012 rather than 2014.

By 2014, however, we are already constrained by the fact that no more than 20 per cent of South Africa's demand can be met by imported hydropower. As a result, South Africa must turn to a gas-fired power station in 2014 (either one on its territory, or one built in Mozambique or Namibia), in lieu of importing more energy from hydropower sources. Indeed, this constraint continues to exercise itself during the subsequent 14 years, with three gas-fired power stations constructed during that period. In 2028, when the share of gas in the region's total production of electricity is at its peak, it stands at 17.8 per cent. With capacity at just under 16 gigawatts (GW), however, we assume that the region has exhausted all of its new potential for gas-fired power stations. Therefore, we enter a third and final phase.

During this phase – 2029 and beyond – the region returns to coal as the main source of its electricity generation. In 2029, South Africa constructs the region's first new coal-fired power station in some 20 years. Though some new hydropower facilities continue to be developed in the Democratic Republic of the Congo (as total demand in South Africa rises, modest imports still enable the total share of imports from hydropower facilities to remain below 20 per cent), so too do coal-fired power stations in South Africa (another four 3000 MW facilities after 2030) and, to a lesser extent, Zimbabwe (one 500 MW plant in each of 2043 and 2049). In 2050, the region's electricity supply profile is as follows: 44.4 per cent coal-fired power stations; 36.7 per cent hydropower; 12.7 per cent gas; 4.9 per cent renewables; 1.5 per cent nuclear; and 0.2 per cent diesel. This compares with baseline figures of: 60.3 per cent coal; 32.6 per cent hydropower; 0.6 per cent gas; 4.8 per cent renewables; 1.5 per cent nuclear; and 0.2 per cent diesel. In the previous hydrophilic scenario, the respective shares were: 34.6 per cent coal; 58.1 per cent hydro; 0.6 per cent gas; 4.9 per cent renewables; 1.5 per cent nuclear; and 0.2 per cent diesel.

In order to calculate both carbon dioxide savings and incremental costs, we focus upon the differences between this scenario and our baseline scenario. For the sake of calculating carbon dioxide savings, the relevant differences

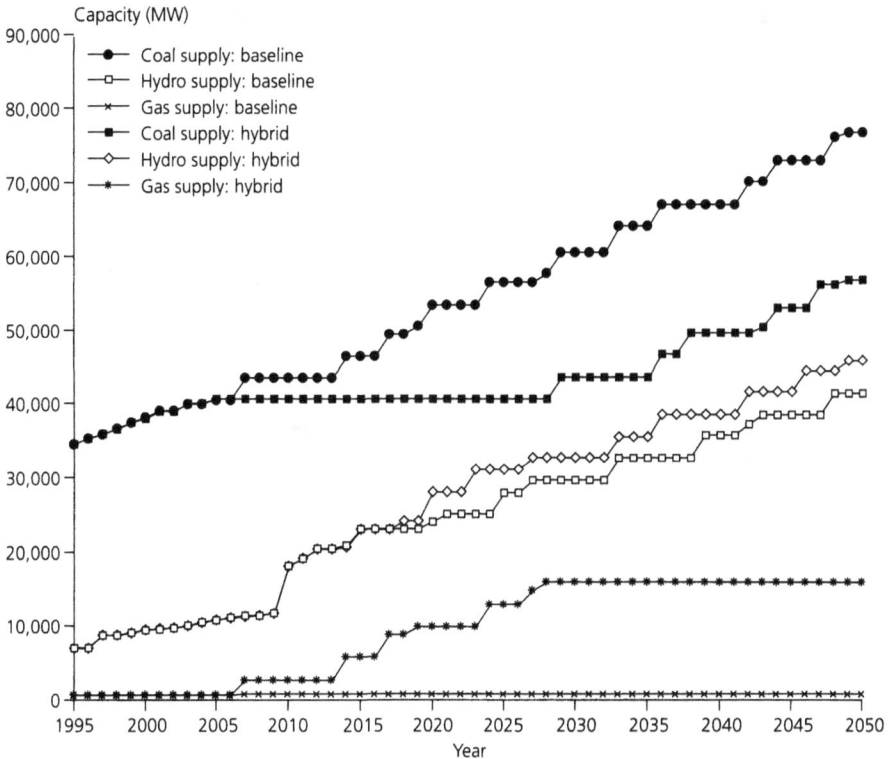

Figure 5.3 *Coal-fired, gas-fired and hydropower capacity in southern Africa, hybrid scenario and baseline*

involve the coal-fired and gas-fired power stations. In this hybrid scenario, six coal-fired power stations are not built in South Africa, nor one in each of Tanzania and Zimbabwe. Moreover, the timing of those that are built is different. In their place, the region has seven new gas-fired power stations.

We calculate emissions from the coal-fired power stations as in the hydrophilic scenario. To calculate emissions from the gas-fired power stations, meanwhile, we assume that these facilities are operated at 35 per cent efficiency, using gas with a carbon content of 56.1 kilogrammes per gigajoule (GJ) (UNEP, 1997, p 102), and have a load factor of 70 per cent. The first year of any reduction is 2007, and the savings amount to 10.8 million tonnes of carbon dioxide. They steadily rise during the subsequent years: 45.7 million tonnes in 2020, 48.4 million tonnes in 2030, and peak between years 2033 and 2037 (and 2042 as well) at a value of 69.2 million tonnes of carbon dioxide a year. In 2050, they amount to 66.3 million tonnes for the year. During the entire study period, savings total 1971 million tonnes of carbon dioxide, just over two-fifths of the savings that we calculated for our hydrophilic scenario. Though smaller, an average annual abatement figure of 45 million tonnes of carbon dioxide is still a consider-

Carbon dioxide emissions (million tonnes)

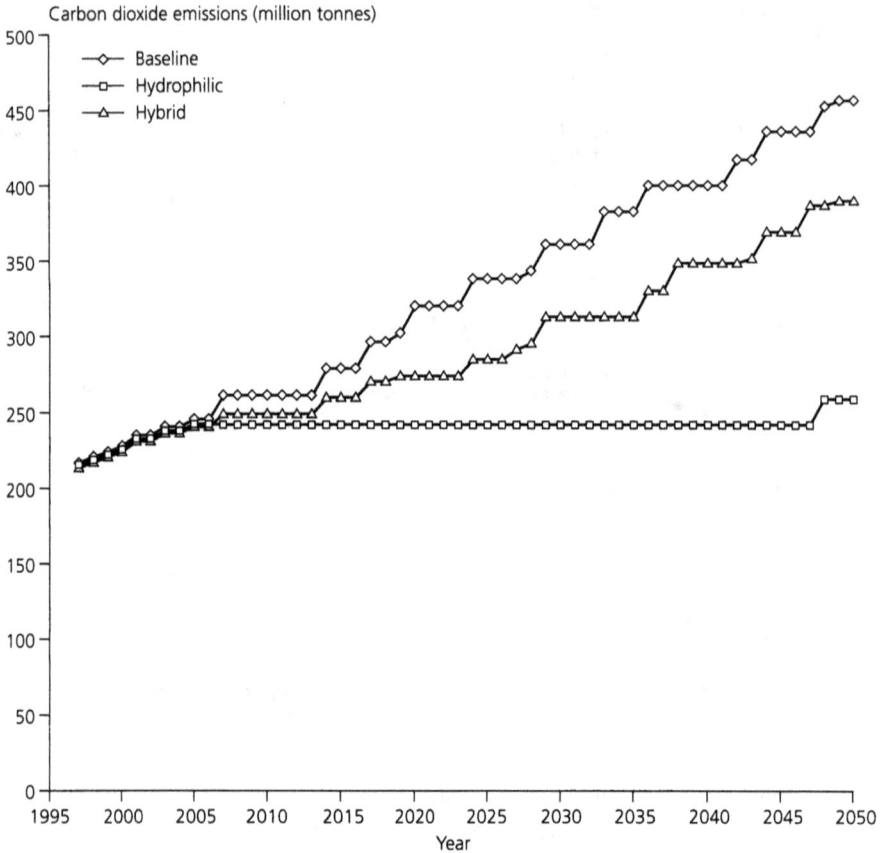

Figure 5.4 *Annual carbon dioxide emissions from the power sector, baseline, hydrophilic and hybrid scenarios, 1997–2050*

able quantity – equal to approximately 10 per cent of Italy's energy-related carbon dioxide emissions in 1992 (World Resources Institute, 1996, p 326). Figure 5.4 shows the total carbon dioxide emissions in each of the three scenarios.

Turning to costs, the new calculation we must undertake is for gas. We assume that the construction of plant will be US$500 per installed kW.[16] Variable costs, meanwhile, consist of first, the fuel and, second, the other annual operating costs (such as staffing and maintenance). For fuel, we estimate the cost of gas to be US$3.30 per gigajoule (GJ) in 1997, rising to US$4.10 per GJ in 2050.[17] For the other annual operating costs, meanwhile, we assume that they will be equal to, annually, US$7.50 per installed kW (which is derived from an estimate of 1.5 per cent of total capital cost).[18]

Calculating the net present value of this scenario – that is, the difference between the net present value of this hybrid scenario and the net present value of the baseline scenario – we discover that the capital costs are smaller in the former (by about 20 per cent). This was certainly to be expected, since the construction costs of gas-fired power stations are considerably lower than the construction costs of coal-fired power stations. The variable costs, however, are higher. Again, this was expected: though the operating and maintenance costs are the same for both gas-fired power stations and coal-fired power stations, fuel costs are considerably higher in the case of the former. Adding the capital and variable costs together, we find that the difference in net present value is US$12.3 billion (with a 5 per cent discount rate). In terms of US dollars per tonne of carbon dioxide abated, this is equivalent to US$6.26.[19]

MITIGATION PROJECTS

By focusing upon the two scenarios in this chapter – that is, the hydrophilic scenario and the hybrid scenario – we are not suggesting that one or the other will be implemented or that they should be the only mitigation options considered. Rather, scenario development, of the kind undertaken in this chapter, can be extremely useful when exploring the implications of a range of activities. Such was our intention.

Nevertheless, because mitigation activity in southern Africa could well develop on a project-by-project basis, we feel that it is also crucial to initiate some analysis of individual mitigation actions in the electricity sector. Here we consider three such possibilities (each of which was also part of one, or both, of our broader mitigation scenarios). Using similar assumptions with respect to carbon dioxide emission factors, loads, efficiency, and costs, we consider three hypothetical projects, each on their own. One is the construction of a 500 MW hydropower facility on the Zambezi River; another is the construction of a 3000 MW hydropower facility in the Democratic Republic of the Congo; and the final one is the construction of a 500 MW gas-fired power plant in Mozambique.[20] For each, we assume that it will displace a coal-fired power station of the same size and that it will come on line in the year 2005 and will operate until 2050. To introduce some sensitivity analysis, we also undertake the same calculations, but only after doubling the capital costs of each of the three mitigation projects. All results are listed in Table 5.4.

Not surprisingly, the two hydropower projects seem to have the greatest potential for carbon dioxide reductions. Moreover, these projects appear to be the cheapest. Again, this is not surprising: we have not costed any potential outages arising from more extensive reliance upon hydropower, nor have we investigated the stability of the resulting system more generally. The third project – the gas-fired power plant in Mozambique – has a higher abatement cost than the broader hybrid scenario we developed in the second part of this chapter. This is due to the fact that the hybrid scenario also made use of a significant quantity of the region's hydropower potential, thereby lowering average costs. The doubling of initial capital costs, meanwhile, has the largest impact upon the hydropower projects. This is a

Table 5.4 *Abatement potential for, and cost of, selected electricity projects in southern Africa*

Project	Carbon dioxide reductions (million tonnes)	Net present value of difference in total costs (millions of US $, assuming 5 per cent discount rate)	Cost of abatement (US $ per tonne carbon dioxide)
500 MW hydropower facility on the Zambezi River	137	−279	−2.04
3000 MW hydropower facility in the DRC	823	−2023	−2.46
500 MW gas-fired power station in Mozambique	56	933	16.71
500 MW hydropower facility on the Zambezi River (capital costs doubled)	137	103	0.75
3000 MW hydropower facility in the DRC (capital costs doubled)	823	70	0.09
500 MW gas-fired power station in Mozambique (capital costs doubled)	56	1,140	20.41

direct consequence of the fact that capital costs make up a larger share of the total costs of hydropower projects (compared to projects involving gas-fired power plants).

Interest in climate change mitigation in southern Africa could well focus upon either broad strategies or narrow activities in the region's electricity sector. We have, in this chapter, aimed to advance discussion about both: firstly, by examining, in depth, two scenarios for the entire power sector in the region; and, secondly, by investigating three representative projects that are contingent upon climate change cooperation in southern Africa.

SUMMARY

The purpose of this chapter has been to investigate mitigation scenarios that depend upon enhanced regional cooperation in electricity supply. After investigating a predominantly hydropower-based scenario (the hydrophilic scenario), we develop a second scenario – one which may make more judicious use of the region's considerable hydropower potential by adding gas to the supply mix. This we call our hybrid scenario.

In both of these scenarios, the potential for carbon dioxide reduction is considerable. In the case of the hydrophilic scenario, emissions for the entire period are over 26 per cent lower than they would be in the baseline scenario. The corresponding figure for the hybrid scenario is 11 per cent. Either way, the quantities involved are significant. Moreover, given that coal-fired power stations are responsible for about one half of South Africa's carbon dioxide emissions (and, by extension, southern Africa's as well), total carbon dioxide reductions are of the order of 5 to 13 per cent.[21] Consequently, it appears that activities in the power sector have the potential for significant carbon dioxide abatement at relatively low cost. What remains for us to consider is the extent to which either scenario – or the individual parts thereof (as we examined towards the end of this chapter) – could be implemented in reality. This is the purpose of the next chapter.

ENDNOTES

1   Full details of the hydrophilic scenario may be found in UNEP (forthcoming).
2   Small exceptions to this are, firstly, the refurbishment of Angola's thermal plants and the development of Sengwa in Zimbabwe (both of which we identified as being part of our utility-derived baseline in the previous chapter). We decide to preserve them in both of our mitigation scenarios, though, of course, it is easy enough to delete them. And secondly, we also retain some new renewable sources of electricity generation after 2030 (which were again part of our baseline from the previous chapter).
3   For comparison's sake, note that Eskom's coal-fired power stations, in 1995, operated at an average efficiency rate of 34.4 per cent and had a load factor of 62.5 per cent (Eskom, 1996a).
4   By examining the *differences* between the two scenarios, we do not need to examine the costs of those elements that are characteristic of both scenarios. Most significantly, these are the costs of replacing or refurbishing older plant in the region.
5   One estimate that has been reported in the press is that the new Majuba coal-fired power station in South Africa will cost Rand 12.5 billion ('Focus on Energy-Efficiency Options', 1997), which is equivalent to approximately US$900 per installed kW.
6   All prices used in this study are in 1997 US dollars.
7   Adapted from UNEP (1997). See also UNEP (forthcoming).
8   A study of future Zimbabwean thermal power stations estimated that the annual cost of operation and maintenance would be '2 per cent of the capital expenditure associated with the plant in service, including expenditure on transmission' (Merz and McLennan, 1981, p 9.20). Given the experience that would be gained in coal-fired power stations, however, we suggest that 1.5 per cent is a more appropriate figure. We also take the US$1000 construction cost per installed kW to come up with the figure quoted here (thus understating, probably, the costs of running the Tanzanian and Zimbabwean facilities; however, the figures for the South African facilities dominate the calculations anyway).
9   We derive this by noting that one estimate suggests that for 'new hydro schemes elsewhere on the Zambezi the annual cost of operation and maintenance is taken to be equal to 1 per cent of the capital cost expended to bring the plant into

commercial service' (Merz and McLennan, 1981, p 9.18). We use a capital cost of US$1000 per installed kW to arrive at the figure quoted here in the text.

10   The choice of an appropriate discount rate could, of course, be the subject of much debate. The *UNEP Guidelines* (UNEP, 1997, pp 30–31) recommend the use of a 'high and a low discount rate for sensitivity analysis ranging from a low 3–5% rate to a high 7–10% rate'. Though, at this stage, we do not undertake sensitivity analyses, we believe that a 5 per cent rate is certainly reasonable.

11   After analysing hydrology in the region, one study concluded that: 'There are strong indications that the present low-flow trend observed over the last twelve years in the Upper Zambezi and Kafue Basins is firmly established and could continue for several years to come...the outlook for the near future calls for a conservative approach to the appraisal of the Kafue and Upper Zambezi catchment yields and hence in the planning for reservoir and hydropower plant operation and for new hydro-installations' (Mukosa, Pitchen and Cadou, 1995, p 97).

12   A recent African Development Bank report – on the prospects for economic integration in southern Africa – noted that:

> Load shedding has a particularly detrimental effect on the productive sector. Before it became clear that the 1992/93 rainy season would be satisfactory [in Zimbabwe], allowing the rationing system to be abandoned, estimates were made of the economic impact of energy curtailment. Even without taking forced load shedding costs into account, electricity curtailment was projected in 1993 to turn GDP growth from positive to negative, involving a reduction of US$500 million (approximately 10 per cent of GDP), lost export revenues of US$435 million (nearly 25% of exports) and the loss of 57,000 jobs or 10 per cent of productive sector employment. (AfDB, 1993b, p 339)

13   This would obviously have consequences for carbon dioxide emissions (existing thermal plant would be working to higher load factors, thus emitting more gases), but we do not examine them explicitly here.

14   For a detailed investigation in another part of the world, see, for example, Beenstock et al (1997).

15   We could also add to this the economic cost of lost investment, since the region might well be perceived as an area with an insecure source of electricity supplies. The social impact of lost employment, or the general sense of insecurity brought about by frequent outages, cannot easily be quantified but should be noted.

16   There are relatively few references to the cost of new gas-fired power stations in southern Africa. One, however, was recently reported by the Economist Intelligence Unit: discussing a possible power plant for the Kudu field gas, it was reported that a 750 MW power station would cost 'US$500 million, comprising US$300 million for the Kudu field development and US$200 million for the plant and links to the existing transmission network' (EIU, 1997, p 19). This is equivalent to US$667 per installed kW (including field development) or US$267 per installed kW (for the plant itself).

17   Adapted from UNEP (1997).

18   'The cost of operation and maintenance for gas turbine power stations containing units of 50 MW installed capacity is assessed at 3 per cent of the cost of the installation' (Merz and McLennan, 1981, p 9.22). We use greater expertise and economies of scale to arrive at our figure of half this amount.

19   Extra costs for outages do not need to be added to this figure since, in every year, the difference between supply and demand is greater than half of the electricity supplied by hydropower plants. Given the three relevant conditions helping to define this scenario – that is, 20 per cent reserve margin, 40 per cent maximum

hydropower contribution and 50 per cent hydropower failure – we could have had, in theory, some outages. A condition that only 33 per cent of the region's electricity (demand plus margin) could be supplied by hydropower would have, under these conditions, ensured no outages. However, given that available supply was always safely above margin (we ensured sufficient margin for each country individually), as well as additional restrictions on South Africa's hydropower imports, we did not, in the end, have to add any costs arising from outages.

20 For the sake of this example, we assume that the capital costs of this hydropower facility are US$1000 per installed kW.

21 Eskom's carbon dioxide emissions in 1995 amounted to 147 million tonnes (Eskom, 1995, p 12), while South Africa's total carbon dioxide emissions have been estimated, for 1992, to be 308 million tonnes (Scholes and van der Merwe, 1996, p 10).

Chapter 6 | Assessing and Implementing Electricity Mitigation Options

R S Maya and Ian H Rowlands

INTRODUCTION

In this chapter, we continue the assessment of the hydrophilic scenario, the hybrid scenario and related projects. Having examined the climate change mitigation potential for each, and the associated financial cost, we turn our attention to other criteria. Following the path laid out in Chapter 1, we consider the broader developmental consequences of each scenario, as well as the prospects for implementing each. In this way, we will be able, towards the end of this chapter, to evaluate the two scenarios more comprehensively by explicitly considering a wide range of potential advantages and disadvantages.

OTHER CRITERIA FOR ASSESSMENT

The areas that we investigate in the rest of this chapter relate either to our interest in the broader developmental consequences or to our interest in the prospects for implementation. We have derived the former by considering the impacts that the scenarios would have upon the region's own developmental ambitions (taken to encompass a range of environmental, economic and social goals), while the latter is based on our discussion in Chapter 2 of this book.

We conclude that there are six key areas for investigation. Each of these highlights particular advantages or disadvantages of each of our two scenarios – in addition to the relative climate change benefits we derived in the previous chapter. In the subsequent six sections, we will explore each with reference to both the hydrophilic scenario and the hybrid scenario.[1]

Technical Issues

Although we have already discussed some of the technical advantages and disadvantages of each of our two mitigation scenarios in Chapter 5, it is worthwhile revisiting the issue here. As a reminder, in the previous chapter we were particularly concerned with the performance of a system that is heavily reliant upon hydropower, given the variable hydrology in southern Africa. We argued that a 50 per cent loss of generating capacity from the region's hydropower facilities was a distinct possibility, and that the costs of such outages must be incorporated into the costing analysis.

We should, however, note that proponents of hydropower argue that its technical performance should also be viewed positively. More specifically, these proponents maintain that hydropower is reliable, based on safe and 'proven technology' (Forsius, 1993, p 9). Additionally, it is able to offer those responsible for a system's operation great flexibility: hydropower's low running costs make it attractive to use for baseload, but its quick start-up time also encourages it to be used for peaking. Moreover, its storage capability makes it a candidate for back-up as well (Garribba, 1993, p 28). Such characteristics encourage Garribba (1993, p 28) to conclude that 'the value of hydroelectricity goes well beyond that of simple capacity and generation'.

The technical strengths and weaknesses obviously have the most implications for our hydrophilic scenario: it is the one that makes the most use of hydropower. In the year 2050 of that scenario, it provides 58 per cent of all electricity generated, compared with 33 per cent in the baseline scenario. Meanwhile, the implications for the hybrid scenario – in which hydropower has a 37 per cent share in the year 2050 – are somewhat less. Indeed, calculations in Chapter 5 have already revealed that we do not need to add any additional costs, owing to the failure of regional hydrology (at least not when following our methodology).

What the hybrid scenario does introduce, however, is gas as a means of electricity generation. This brings with it a couple of technical benefits. The first is that gas-fired power stations offer great flexibility for both long-term planning and short-term performance. In the case of the former, because gas-fired power stations can be constructed relatively quickly, they provide decision-makers with considerable freedom of implementation. In the case of the latter, because these stations can be fired up relatively quickly, they are effective for operation as either peaking or baseload.

A second advantage is that introducing gas into southern Africa's electricity future increases the range (both geological and geographical) of regional resources that are being used for electricity generation. This is advantageous because supply centres will be more widely dispersed (thus enhancing reliability), and there will be less dependence upon individual large power-generating facilities. Moreover, other uses for gas – perhaps industrial or household – could be encouraged.

Local Environmental, Economic and Social Impacts

Though hydropower is significant in both of our scenarios, it is particularly dominant in our hydrophilic scenario; as noted above, in the year 2050 of that scenario, it provides 58 per cent of all electricity generated, compared with almost 37 per cent in the hybrid scenario and 33 per cent in the baseline scenario. We must recognize, however, that hydropower not only serves to generate electricity, but controversy as well. This has significant implications for any assessment of our scenarios.

In brief, some critics charge that hydropower creates a range of environmental, economic and social problems which outweigh any resultant benefits. Full discussion of this broader debate about hydropower is beyond the scope of this chapter. The reader is directed to writings by both opponents and proponents in this debate.[2]

Let us consider some of the potential local impacts of greater use of hydropower. We begin by focusing upon the hydrophilic scenario since it is the one that makes the most extensive use of the region's hydropower resources. Clearly, a comprehensive analysis would involve a thorough impact assessment of each site that is both exploited in the hydrophilic scenario and left unused in the baseline scenario. This would basically involve new locations generating 7500 MW in Angola and 27,000 MW in the Democratic Republic of the Congo.[3] Given the relatively small amount of work completed on Angola's hydropower potential – indeed, in our scenario, we have not even identified specific sites for these new developments – detailed impact assessments are unavailable. Much of the analysis of the potential Inga project, meanwhile, has concentrated upon the impact of transmitting electricity (particularly to South Africa and Egypt) rather than generating it (see, for example, 'The Great Inga Dream', 1995). Even these studies, however, are not readily available in the public domain. As a result, we launch a more general discussion here, striving to discover how each site's particular characteristics allow us to estimate, broadly, the local impacts. Though this is no replacement for a detailed environmental, economic and social assessment, it is an effective way of providing an initial order of magnitude estimate and encouraging further debate on the subject.

The large Inga hydroelectric project in the Democratic Republic of the Congo would be a 'run-of-river' facility.[4] As a consequence, many of the most deleterious environmental and social problems associated with traditional reservoir dams could well be avoided (Goodland, 1995). In particular, because a relatively small area would be flooded, there may be less of an impact upon the local environment (including the local climate) and the indigenous peoples of the area; potential damaging impacts upon the global climate, moreover, could also be minimized.[5]

Inga's 'run-of-river' status, however, does not necessarily exonerate it: Usher, for example, argues that three such projects investigated in her edited volume still generated considerable environmental and social problems (Usher, 1997, p 9 and passim). Nevertheless, the fact that Inga is a run-of-river project should increase the chances that its harmful environmental and social impacts are minimized. All that is certain at this stage is that these southern African hydroelectric projects should not be rejected out-of-hand;

instead, a more rigorous examination of their environmental and social consequences should be undertaken.

Turning to the local economic impact, hydropower facilities have often placed significant stresses upon the host country's financial resources. Indeed, not only are the estimated capital costs of hydropower facilities considerable, but the prospects for cost overrun also appear to be large. A World Bank study of 66 hydroelectric plants approved for financing by the bank between 1965 and 1986 found that cost overruns averaged about 27 per cent (Bacon et al, 1996, p 29); that was even after five hydropower project outliers had been rejected.[6] Hydropower projects can take longer than anticipated to build as well: 'forty-nine hydroprojects reviewed by the World Bank's Industry and Energy Department in 1990 took on average five years and eight months to build, fourteen months longer than the average pre-construction estimate' (quoted in McCully, 1996, p 270).[7] More broadly, meanwhile, there are also concerns that any particular hydropower project may have negative macroeconomic impacts upon the country: depending upon the particular financing scheme employed, a single hydropower project could absorb a country's entire development budget for many years. As a result, other important national projects could be crowded out (Besant-Jones, 1995).

The worst of these impacts is likely to be avoided in southern Africa – indeed, the worst of these impacts will probably be avoided virtually everywhere in the world. The reason for this is that the private sector is increasingly replacing the public sector as the main financing agent of hydropower projects. The consequences of this for ensuring that sufficient financial resources will be available are examined below. For our purposes, however, it is more important to recognize that a large national debt is less likely to result from new hydropower projects today, compared with ten or 20 years ago. This lends further support to our tentative affirmation that such projects – and, indeed, the broader hydrophilic scenario – should not be rejected solely on the basis of speculative local impacts.

Moreover, the developmental consequences of the activities that these mitigation options would either displace or delay are also worth thinking about. In both cases, we are looking at coal-fired power generation which, in terms of life-cycle analysis, is the most damaging of the fossil fuels (UN, 1994). One study found that the environmental externalities of a conventional coal-fired condensing plant were valued at between 2.5 and 39 mecu per kWh (approximately 0.27 to 4.28 US cents per kWh) (Meyer et al, 1996, p 25). Indeed in the particular case of South Africa, van Horen (1996) has quantified the external costs of coal-generated electricity, arriving at a range of between 0.6 and 3.4 US cents per kWh.[8] Of the various impacts he examines, the health consequences of air pollution from power stations receive considerable attention. Van Horen (1996, p 77) concludes that factors like asthma attacks, respiratory symptom days and days of restricted activity are significant. Of course, there are also positive developmental consequences of coal-fired power stations; the employment that they provide is perhaps amongst the most significant. Some governments or societies may rate such social externalities quite highly (as the British government has in the past).

We have been concentrating upon hydropower up to this point. It is, however, also important to consider the introduction of gas (which is part of the hybrid scenario). Consequently, a few words about the local environmental, economic and social consequences of gas are certainly in order.

Though by no means benign, gas is often characterized as the most environmentally friendly fossil fuel. Indeed, its lesser impact upon the global climate encouraged us to include it in our hybrid scenario. Moreover, its smaller impact upon local environments can also be viewed as positive: on combustion, gas emits less air pollution than other fossil fuels. Problems may arise, however, in the transportation of gas. There is 'high leak potential because of high pressures and corrosion' (UN, 1994, p 42), which could result in explosions. Moreover, leakages also exacerbate global climate change: methane is an extremely powerful greenhouse gas.[9] Consequently, excessive leaking could serve to negate any climate gains that resulted from reduced carbon dioxide emissions. This could, however, be avoided by transporting the electricity generated by the gas, rather than the gas itself. As in the case of hydropower projects, this warrants much closer attention.

The capture and use of coal-bed methane, as opposed to natural gas in reservoirs, generates more developmental uncertainty. A 1995 study for SADC concluded that the production of coal-bed methane:

> ...contains more environmental problems than natural gas. This relates to land-use issues due to the need for a large number of production wells and in-field lines and the production of waste water with dissolved solid materials. However, the following characteristics of CBM utilization may more than compensate for these problems:
> (1) The use of CBM utilizes a resource which was previously allowed to escape. Therefore the efficiency of use of natural resources has increased.[10]
> (2) The de-gassing of coal before mining of coal will increase safety factors by minimizing possibilities for methane explosions in coal mines.
> (SADC, 1995c, p 3.1)

Inevitably, virtually any activity will generate some kind of deleterious local consequences. This section, therefore, has tried to identify some of the local environmental, economic and social consequences of our two scenarios – particularly by concentrating upon the greater use that they make of hydropower and gas. Given the incompleteness of the picture that emerges, more work in this area is clearly needed. Nevertheless, results are sufficiently encouraging as to oppose immediate rejection of our two mitigation scenarios. Instead, we continue to consider different developmental and implementation issues.

Security of Supply

What drove the development of both of these scenarios was the desire to lessen the net emissions of carbon dioxide, which in turn was realized by an increase in the level of regional electricity exchanges. As a reminder, Table 6.1 lays out the percentage of electricity generated that was imported from international markets in each of the three scenarios (for selected years), while Table 6.2 lists similar figures for South African imports in particular.

From these, it is clear that the increased internationalization of power is a key characteristic of each scenario. In our hydrophilic scenario, we find that, of the large economies in the region, three of them import substantial amounts of their electricity needs. South Africa stands out in this regard. By relying upon almost 41,000 MW of 'foreign' capacity in year 2050, approximately 47 per cent of its demand (including reserve margin) is satisfied by electricity generated outside of its national borders. The corresponding figures for Tanzania and Zimbabwe are 48 per cent and 45 per cent respectively. Indeed, even though an effort to ameliorate imports is explicitly built into our hybrid scenario – in the form of the 20 per cent cap that South Africa has upon its imports of hydropower – significant international transfers of electricity still take place.

Table 6.1 *Percentage of electricity in southern Africa supplied by another country*

	2000	2010	2020	2030	2040	2050
Baseline	6.8	10.1	12.5	14.5	16.3	16.2
Hydrophilic	6.8	15.3	25.5	34.5	39.7	43.2
Hybrid	6.7	10.6	20.7	24.9	25.2	27.1

Sources: Chapters 4 and 5; and UNEP (forthcoming)

Table 6.2 *Percentage of South African electricity supply that is imported*

	2000	2010	2020	2030	2040	2050
Baseline	4.1	10.9	12.4	13.8	14.5	12.5
Hydrophilic	4.1	17.5	30.2	39.9	46.3	46.7
Hybrid	4.1	10.9	24.0	28.3	27.1	27.9

Sources: Chapters 4 and 5; and UNEP (forthcoming)

One of the key challenges to implementing either of our mitigation scenarios would be the need to overcome some national leaders' natural resistance to relying heavily upon imported electricity (or indeed the importation of any goods that were deemed in some way to be strategic). In Chapter 2, we have already observed that concern about the ceding of sovereignty is often an impediment to regional cooperative efforts, generally. More specifically related to our issue, a country would rather generate its own electricity, all else being equal, than import it. This is because electricity (and energy more generally) is seen to be a crucial contributor to a country's overall prosperity, and thus it is important to place its production in reliable hands. Given the fact that sovereignty rests in the level of the nation state, insecurity is often seen to increase once an entity in another sovereign nation state is entrusted with the task of generating one's own electricity. As a result, some countries would prefer to pay somewhat more for generating their own electricity than allowing others to generate it for them – indeed, such a risk premium, placed at 15 per cent, was quoted by the SADC TAU's *Manual for Economic Analysis of Energy Projects*. Though greater stability in southern Africa during the 1990s may well have reduced the exact level of this premium, it could nevertheless still exist (either consciously or unconsciously).

As part of any power-sharing arrangement, some degree of international trust – if not a formal transfer of sovereignty – will be crucial. Røde et al (1995, p 4), for example, argue that an actual power pool agreement 'reduces the sovereignty of a member with regard to the installation, operation and maintenance of equipment'. They go on to note that '[because] the utilities tend to be closely linked to their respective governments, the involvement of governments is unavoidable. The decisions of some members could therefore be dictated by politics and not by economics' (Røde et al, 1995, p 4). Thus, effective implementation could be hindered by issues related to sovereignty; whether for reasons of security of supply or political expediency, national leaders may resist greater regional cooperation on electricity supply. Indeed, the Appendix to Chapter 3 revealed how desires for sovereign control over electricity supply in southern Africa have already supplanted regional schemes with national projects.

How might leaders' concerns about the potential loss of sovereignty be ameliorated? And how might we enhance the prospects for successfully implementing a regional mitigation strategy? A few possibilities exist. Firstly, diversity of supply is important. While there are real economic attractions to developing a particular electricity supply site to its limits (economies of scale in plant-size construction, full use of transmission lines and so on), there are significant benefits from having a range of supply locations (especially if they are in different countries, or at least have different owners). Purchasing bodies will be more willing to rely upon extraterritorial generation if they know that their suppliers are located in many different countries since this will help to ensure more competitive pricing (through freer competition), as well as enhanced reliability (through greater number of suppliers). The level of anxiety among national leaders would inevitably heighten should only one source of electricity dominate.

Diversity of supply characterizes both of our mitigation scenarios. In 2050, we envisage that South Africa (the main importer) will purchase electricity generated by hydropower facilities in the Democratic Republic of the Congo, Mozambique and Angola, and perhaps even a little in Zambia as well. Moreover, this list consists of countries located in different parts of the region, allowing their electricity to be delivered by completely different transmission lines (not only do we have to consider the market for electricity generation, but also for electricity transmission, or 'wheeling'). In addition, in our hybrid scenario, South Africa will import electricity generated by gas from Namibia and Mozambique, in addition to the aforementioned hydropower sources. This extremely diverse supply bodes well for successful implementation.

What matters is not only the fact that there are different countries supplying electricity, but also the particular condition of those different countries – that is, national characteristics such as level of development, political stability and so on. Relying on three countries for electricity supply may not necessarily be better than relying on two, particularly if the three are the poorest and least stable in the region. Though we cannot necessarily predict which southern African countries will score higher or lower on these measures during the next 50 years, we can argue that promoting peace and prosperity in the region, generally, could encourage effective implementation of virtually any regional mitigation scenario.

Organizational development, generally, would also help to lessen worries about loss of sovereignty. If the region's policy-makers were to meet regularly within a transparent institutional framework, then their confidence in regional activities would surely increase. Not only has the Southern African Power Pool (SAPP) experience been indicative (see the Appendix to Chapter 3 for a more detailed discussion), but earlier bodies – such as the SADC Energy Sector Technical and Administrative Unit Electricity Section and the SADC Electricity Sector Subcommittee (which is one of the most successful and effective subcommittees in the SADC Energy Sector) – also helped to increase the level of trust among the region's utility officials. With the advent of the SAPP – in particular, its protocols and memoranda, as well as its committee and subcommittee structures – we are seeing the further institutionalization of regional networks. While this does not suggest that more institutions are necessarily better, we certainly imply that some kind of organizational development could help to implement scenarios that depend on increased regional cooperation.[11]

The ability for an entity in one country to purchase capacity (generating or transmission) in another country would also help to alleviate concerns about the ceding of sovereignty. An entity will clearly have more control over outcomes in another country when it owns and operates the facilities concerned. Indeed, we have already seen movement towards such extraterritorial ownership in southern Africa. In mid 1997, for example, Eskom (the South African utility) bid for some of the generators previously owned by the Zambian Consolidated Copper Mines (ZCCM).

> *Because of its location, the ZCCM system could play a pivotal role in the supply of electricity throughout Southern Africa, says Eskom media relations manager Peter Adams. Gaining control of the system could allow Eskom to influence the development of the Southern Africa power grid to a considerable degree.* (Engineering News, 1997)

This will, of course, not be a cure-all for anxiety about foreign generating capacity, but it would certainly help.[12]

In addition to extraterritorial activity by governmental or quasi-governmental organizations, we consider similar activity by non-governmental organizations. In this regard, the growing importance of independent power producers (IPPs) is worthy of note, since it represents a new and significant regional trend in southern Africa. Leading in this respect are Zimbabwe, where independent producers (YTL of Malaysia and National Power of the United Kingdom) have entered into governmental agreements to construct additional generating capacity, and Zambia, where a possible independent grid-operating company is being considered.

The fact that IPPs are major players in southern African electricity supply will undoubtedly have significant consequences. For one, states may resist relinquishing control to IPPs (just as they oppose ceding authority to other states, or any kind of supranational authority). This is not surprising, given that the history of the power sector in most southern African states involves only parastatal utilities and municipal utilities, with only lesser participation by private generators (primarily running small diesel plants at remote locations). Indeed, the fact that membership of the Southern African Power Pool has, thus far, remained limited to state utilities may be indicative. Though the reasons for excluding IPPs (and municipal utilities) are unclear, there nevertheless appears to be a conscious effort to keep non-state utilities outside of the process. Consequently, although the participation of IPPs may be crucial (for financial reasons, which are elaborated below), the fact that national utilities (and national governments) may see them as threats may hinder implementation. This, however, could equally be said of our baseline scenario as it could of either mitigation scenario.

Nevertheless, while national sovereignty should be considered as a potential hurdle for implementing any mitigation option, two factors are likely to weigh heavily against it exercising overwhelming influence: firstly, liberalization of the power sector is proceeding, with IPPs admitted into what has historically been the domain of state parastatals; and secondly, there are strong technical and economic motivations for the region's power utilities to cooperate more fully. Southern Africa – a region which has witnessed successful regional economic cooperation in the past – should be able to muster the will and resources to overcome traditional concerns about ceding sovereignty.

Distribution of Benefits and Costs

Recalling Chapter 2, we can also think about how the interests of different countries are affected by increased electricity sharing. Though regional action in electricity generation may create benefits for the countries taken together as a whole, there may be an unequal distribution of those benefits. In contrast with the expectations generated by the discussion in Chapter 2, however, Røde et al (1995, p 4) argue that it will not inevitably be a case of polarization: 'In a pool, the benefits to a member are seldom proportional to the support given by that member to the pool. The larger utilities invariably give larger financial and technical support to the pool, while the smaller utilities reap the proportionally larger benefits.'

The fundamental characteristic of each mitigation scenario is not only that those countries which import electricity in the baseline scenario will import more, but the corollary as well – namely, those that export in the baseline scenario will export even more. Angola and the Democratic Republic of the Congo (DRC) are primary examples of this. Rising exports could generate significant earnings for these countries. The African Development Bank, in 1993, reported that: 'On a conservative estimate, 8,000 MW of export capacity by 2010 could generate a net income for supplying countries of US$800 million annually' (AfDB, 1993a, p 19). In our hydrophilic scenario, Angola's exports peak at 9500 MW during the 2040s (compared with 2000 MW in the baseline), while the DRC's rise to 36,000 MW during the same decade (compared with 12,000 MW in the baseline). Therefore, the scenario could well be worth approximately US$750 million a year to Angola by the end of the period, and over US$2.5 billion a year to the DRC at the same time. Consequently, interest on their part in the hydrophilic scenario should not come as a surprise.[13]

Countries that generate electricity accrue revenue, but so too do those that transport it from the power plant to the demand centres. In our hydrophilic scenario, both Angola and Namibia would certainly be important 'wheeling countries' (countries across which electricity flows).[14] Zambia, Zimbabwe and Botswana could also prove to be the same, particularly if South Africa encouraged a diversity of supply routes (for the reasons outlined above).

Wherever there are exporters, there are also importers. We have already noted how South Africa's imports rise in our hydrophilic scenario (peaking at 40,950 MW a year, in the second half of the 2040s, compared with 10,950 MW in the baseline scenario). Zimbabwe and Tanzania are the second and third largest importers of electricity in the region: in the case of the former, 5000 MW by the end of the study period (compared with 2500 MW in the baseline), and in the case of the latter, 4000 MW (compared with 3000 MW in the baseline). Though the net cost of electricity to these countries might not be higher, the particular distribution of costs and benefits may well encourage factions within these countries to resist a 'hydrophilic' future. Though this may hinder implementation, others may welcome its redistributive impact. Let us expand upon this point.

Were Angola and the DRC to build significant hydropower export plants – and countries such as Zambia, Zimbabwe and Botswana to transport the electricity generated – a massive transfer of funds from the importing countries (primarily South Africa) to the exporting and wheeling countries would result. This should, we maintain, be a positive development for the region as a whole since it would redress the large trade imbalances that currently exist among southern African countries. South Africa, in particular, is a dominant export country; in return, however, it imports relatively little from the other states in the region.[15] Electricity exports would therefore help to reduce the tension associated with trade imbalances that are currently being experienced. This particular distribution of benefits in the region arising from the hydrophilic scenario could advance other regional goals (particularly those of equity and stability).

The extent of redistribution is more modest in the hybrid scenario: Angola's exports rise only to 3500 MW by 2049, and the DRC's to 15,000 MW. Nevertheless, export earnings could increase by US$150 million for the former and US$300 million for the latter. Clearly, that would still be significant. Similarly, imports on the part of South Africa do not rise as much (from 10,950 MW to 24,950 MW in the year 2050), while they do not rise at all for either Zimbabwe or Tanzania. Accordingly, though the above effects still apply, they do so to a lesser extent.

In summary, the distribution of costs and benefits is potentially an overriding impediment to power-sector cooperation among southern African countries. This subject needs to be studied in more detail so that the benefits for each member state and participating utility are clearly outlined.

Availability of Finance

Financing the electricity supply industry in the developing world will inevitably be a challenge during the years to come. Consequently, it could be a barrier to successfully implementing any future scenario – our baseline scenario included![16] However, what we want to explore in this section is the extent to which introducing our two mitigation scenarios poses additional barriers to financing, above and beyond those thrown up by the baseline scenario (or, alternatively, the extent to which introducing our two mitigation scenarios reduces barriers present in the baseline scenario, or even encourages successful implementation in their own right).

It is important to recognize that the cost of each of our two mitigation scenarios consists of two elements – the baseline cost (an amount that is equal to the cost of the baseline scenario) and the incremental cost (an amount that is in addition to the cost of the baseline scenario). As noted in

Chapter 1 of this book, we assume that the incremental cost may well be paid by an international organization or mechanism because of the global climate gains that would arise from its successful implementation. One may therefore believe that, given that the hitherto unpaid baseline cost is the same for both the baseline scenario and the mitigation scenario, there would be no additional financial challenge (the same amount of money must be sourced for both activities). Indeed, given that the incremental cost is assumed to come from the international community, one may well wonder why our mitigation scenarios would pose a financing challenge of any sort.

We maintain that our mitigation scenarios introduce new financing issues, for two reasons. Firstly, the mitigation scenarios use different natural resources, and secondly, they depend upon regional cooperation to a greater extent. We will consider each of these in turn.

It is our contention that the ease with which X dollars can be found to pay for a coal-fired power plant is not the same as the ease with which X dollars can be found to pay for a hydropower station (nor, for that matter, is it the same as the ease with which X dollars can be found to pay for a gas-fired power station). Because the type of natural resource matters, we consider the financing implications of using different resources. To begin with, we concentrate on the implications of having a significant measure of hydropower in each of our two mitigation scenarios. Traditionally, many hydropower projects have been financed with public money. Given, however, changes in prevailing financial profiles (particularly those of utilities and governments in the developing world),[17] many believe that resources will either have to be in the form of grants from international or extranational bodies, or international capital attracted to the project on strict profitability criteria.[18]

What will affect, however, the attitude of international financiers – whether they are governmental, intergovernmental or non-governmental – is the broader debate surrounding hydroelectric projects. This has been outlined above. For our purposes, it is important to recognize that, given the substantial criticism levelled against large dams – in particular, the discussions surrounding the Narmada project in India and the Three Gorges facility in China – this 'global discourse' will certainly affect the prospects for introducing additional hydropower into southern Africa (which amounts to 31,500 MW in the case of our hydrophilic scenario).

The development of any single dam in the region would inevitably attract international attention, and at least some criticism.[19] Indeed, since a number of non-governmental organizations have called for an 'immediate moratorium on the building of large dams' (until a number of conditions have been satisfied), opposition would be virtually guaranteed.[20] This could well affect financing for hydropower projects in the region: domestic lobbying of either public or private financing institutions could sway decisions over allocation. Any potential foreign investor, therefore, might well attach some kind of 'bad press premium' to the project – if it is not extremely attractive, then they will look elsewhere for a non-hydropower activity. Let us examine, more closely, some of the key potential investors.

For governmental and intergovernmental entities, the future of hydropower financing may be most affected by a recently formed independent international dam review commission. Together, the World Conservation Union (IUCN) and the World Bank have been investigating the most appropriate role for large dams in the development process. They have agreed to form a World Commission, which has the following terms of reference (Dorcey, 1997, pp 9–10):

- to assess the experience with existing, new and proposed large dam projects in order to improve (existing) practices and social and environmental conditions;
- to develop decision-making criteria and policy and regulatory frameworks for assessing alternatives for energy and water resources development;
- to evaluate the development effectiveness of large dams;
- to develop and promote internationally acceptable standards for the planning, assessment, design, construction, operation and monitoring of large dam projects and, if the dams are built, to ensure affected peoples are better off;
- to identify the implications for institutional, policy and financial arrangements so that benefits, costs and risks are equitably shared at the global, national and local levels; and
- to recommend interim modifications, where necessary, of existing policies and guidelines, and to promote best practices.

Clearly, the outcome of these discussions (which are scheduled to last from late 1997 to late 1999) will be highly influential for the future of hydropower projects the world over, southern African included.

Turning to the private sector, where entities are more likely to concentrate upon the 'bottom line', many have argued that they will probably not be interested in hydropower projects. Capturing this sentiment, McCully reports on a recent conference:

> An air of gloomy resignation pervaded the meeting. Several speakers from the financing side of the industry emphasized that private investors are discouraged from backing hydrodams because of their high initial construction costs, long capital payback periods, terrible record of construction time and cost overruns, and high operating risks, especially because of their vulnerability to drought. The speakers also made clear that financiers are dissuaded from funding dams because of 'environmental risks': delays because of opposition to resettlement and anti-dams campaigns, and new environmental legislation to regulate how dams are built and operated...Many papers from this conference generally concluded that the funding for hydro that probably will

> *take place will be for small and medium run-of-river dams.* (McCully, 1996, p 274; see, also, Majot, 1996)

While this does not bode particularly well for private-sector participation in hydropower financing, a more general survey of investment prospects in the region suggests that all might not be so gloomy.

Recently, much international attention has focused upon the foreign investment potential of southern Africa. Conferences sponsored by, for example, the World Economic Forum[21] and the *International Herald Tribune*[22] have brought together government officials, businesspeople and others from both within the region and beyond. Though it is by no means certain that any kind of foreign investment will be forthcoming (let alone foreign investment that advances sustainable development in the region), many believe that its arrival is much more likely now than it was just a few years ago. What with Africa being seen by some as the 'last frontier of business', the financing of energy projects – which are a key part of the region's infrastructural development – could ride this larger wave.[23]

Clearly, a new set of questions for financing are posed when the resource involved is hydropower. Indeed, Besant-Jones (1996) calls it 'the distinctive issue for hydro IPPs'. While much of the literature focuses upon the needs of the potential investor (for example, Besant-Jones, 1996), close attention must also be paid to the needs of the region. Any offer of foreign investment that is forthcoming must, of course, advance the region's own developmental goals as well. Meeting the demands of the international investor is only part of the challenge.

Once again, we have concentrated upon the hydropower elements. This is only natural given that it is a significant part of all three of our scenarios. However, since gas is also introduced in one, we now turn our attention there. Particular characteristics of gas that might affect its financing prospects are, firstly, its positive image in terms of its lower impact upon the environment, both global and local. Indeed, the World Bank appears to be quite bullish on gas: 'The World Bank's new lending policy to the oil and gas sector emphasizes gas over oil and infrastructure rather than upstream activity. Promoting gas pipeline transmission projects is very much in consonance with this new emphasis' (World Bank IENOG, 1996). This suggests that implementation would be supported by international players.

The fact that our mitigation proposals require more extensive regional cooperation has significant implications for financing prospects. More specifically, because of the requirement to engage more than one country (and hence more than one government, more than one legal system, and so on) in any agreement, the complexity of the requisite financing arrangement will certainly increase.[24] Accordingly, transaction costs for the entire endeavour will also be higher. Therefore, financing challenges arise by virtue of the kind of resource exploited and the international nature of the endeavour.

Pricing

Homogeneity was identified in Chapter 2 as a key factor in the prospects for regional cooperation. In the case of power sharing, homogeneity is required in many areas. For instance, there needs to be harmony among the different national pricing systems; Charpentier and Schenk (1995, p 3) argue that 'only a pricing system based on bidding is viable because it does not require the publication and verification of detailed economic information'. Indeed, pricing issues over electricity have often proved difficult to resolve when they have involved two or more countries in the region. The classic example involves Botswana, Zimbabwe and Zambia.

After the commissioning of the SADC Interconnector (see also the discussion in the Appendix to Chapter 3) between Botswana and Zimbabwe in June 1990, electricity failed to flow for another nine months. During this time, agreement on tariffs could not be reached. 'Discussions among representatives were fervent, as each tried to secure the best deal possible' (Rowlands, 1994, p 137; see also Jordanger, 1992, p 15). Although agreement was eventually reached, the lesson was clear: tariff negotiation will not necessarily be an easy process. Moreover, the fact that Zimbabwe has not always paid Zambia for electricity purchases on time ('Zambia Needs More Time', 1993, p 10), and that similar tensions have arisen between Mozambique and South Africa, lend further weight to this message.

Indeed, the pricing of electricity continues to be one of the major issues facing southern African utilities. Traditionally, government has controlled electricity pricing: utilities applied to governments for permission to increase prices. Because there was no specific formula for electricity pricing, prices would not necessarily reflect supply costs nor would they necessarily engender efficiency in production. Of the 12 national power utilities in the region, only five can be said to charge cost-reflective tariffs. There is also a significant amount of cross-subsidization which, when coupled with taxes on consumers, adds to the distortion of prices.

With increasing liberalization of the sector, some utilities have opted for long-run marginal cost pricing. However, even this approach would still be made ineffective by the fact that the bulk of the electricity is supplied, in some countries, via coal which is not on long-run marginal cost pricing, but rather on cost-plus pricing formula; this, in turn, distorts production costs and encourages inefficiencies.

If our mitigation scenarios were to be implemented under a loose interconnector arrangement (as is presently the case), then the price of international electricity sales would probably continue to be set by means of contract pricing (as is presently the case). If, alternatively, our scenarios were to be realized by means of a more formalized power pool, then pricing mechanisms could become much more intricate. For example, a competitive bidding system could be set up where utilities could bid daily to supply a purchaser's needs for a competitive price. Regardless of what develops, the point is simply that a pricing regime that is acceptable to all will be a crucial prerequisite for successfully implementing either mitigation scenario.

Power pricing is by far the most problematic aspect of national electricity-sector management in most southern African countries. It is a complicated issue, which must be handled in a sensitive manner for at least three reasons: firstly, price changes have the potential to destabilize small uneconomic utilities in the region; secondly, they have the potential to upset strategic industries that are dependent on high electricity consumption; and thirdly, they have the potential to generate damaging social consequences, particularly during the transition from subsidized electricity pricing to full-cost pricing. Extending pricing issues to the regional level can only serve to augment the complications.

TOTAL ASSESSMENT

Our purpose in this chapter was to widen the criteria for assessment – beyond simply the economic cost of mitigation – to include broader developmental impacts, as well as prospects for implementation. Having considered this, we now review our findings and tentatively take stock of our position. To do that, we have gathered what we consider to be the main advantages and disadvantages of the two scenarios that we have developed. They are listed in Table 6.3 below.

Not surprisingly, although the advantages accruing from the hybrid scenario are generally not as large as those arising from the hydrophilic scenario, neither are the disadvantages. Consequently, the decisive criteria for selecting particular elements of either may be the quantity of abatement desired, and what resources might be used to dampen the disadvantages. We have made no attempt to value the advantages and disadvantages here, in order to decide which (if either) of the scenarios is preferable. Because one of the primary aims of this study is to develop methodological ideas, we leave them simply in table form. People will obviously value different elements in different manners.

What is clear is that we could conceive of situations in which either of the two scenarios are preferable. If a high premium were placed upon carbon dioxide abatement, then the hydrophilic scenario could well be worth pursuing. If carbon dioxide abatement were still a goal, but not one that towered above all others, then the hybrid scenario may be deemed most desirable by society. Moreover, given that these are only two of literally thousands of possible scenarios – as the development of individual mitigation projects at the end of Chapter 5 reminded us – others will inevitably be most appropriate in other situations.

Table 6.3 *The relative advantages and disadvantages of the two mitigation scenarios*

Hydrophilic Scenario		Hybrid Scenario	
Advantages	Disadvantages	Advantages	Disadvantages
• large quantity of carbon dioxide abated • considerable intra-regional financial transfers, which could promote peace, development and equity in the region • mitigates damaging local environmental problems arising from coal mining and combustion	• small likelihood that South Africa would be willing to rely so heavily upon imports • heavily dependent upon the region's fragile hydrology • extensive dam construction and operation brings concomitant environmental, economic and social problems	• considerable quantity of carbon dioxide abated • mitigates damaging local environmental problems arising from coal mining and combustion • develops a wider range of energy-supply industries in southern Africa • the gas elements of the scenario appear to be more easily fundable	• South Africa may hesitate to rely so heavily upon imports • dam construction and operation may generate environmental, economic and social problems

SUMMARY

The purpose of this chapter has been to launch a wider assessment of electricity mitigation options that depend upon enhanced regional cooperation in southern Africa. To structure the investigation, we continued to use the two broad scenarios that we developed in the previous chapter. Nevertheless, many of our comments apply to the individual projects that constitute each (or sometimes both) of those two scenarios.

A general conclusion that emerges from this chapter is that those mitigation activities that correlate well with the region's own development priorities will have the best chance of successful implementation, and therefore will be the most likely to end up contributing to global climate change mitigation. Put in slightly different terms, those options that are of most interest to southern Africans, and those that involve southern African participation and ownership to the greatest extent, are the ones that are most likely to advance global climate change ambitions successfully. By means of examining a number of potential advantages and disadvantages in this chapter, we have seen that three elements – climate change mitigation, developmental assessment and implementation prospects – are closely linked; achievement of the first, by means of the third, requires a reading on the second that is positive. This will generate true win–win possibilities for both the global and regional communities.

ENDNOTES

1   Though many of our comments draw attention to either or both of our two broad mitigation scenarios, most would be equally relevant to any discussion of either individual hydropower projects or individual gas-fired power projects – like the mitigation projects discussed at the end of Chapter 5.
2   In the case of the former, see Goldsmith and Hildyard, 1984; and McCully, 1996. For examples of the latter, see Forsius, 1993; ICOLD, 1997; and Seabra, 1993. Dorcey, 1997; Moreira and Poole, 1993; and WEC, 1995b; meanwhile, provide broader overviews.
3   Other differences between the two scenarios are the bringing forward in time of an additional 20,240 MW of hydropower capacity. Given that the associated local impacts of these projects will be felt for a longer period of time in our hydrophilic scenario (compared with our baseline scenario), we might want to examine them more closely as well. For now, however, we concentrate upon the new locations exploited in the hydrophilic scenario.
4   By run-of-river, we mean that the flow in the Congo River is sufficiently strong and steady that only very limited storage capacity is necessary.
5   With regard to contributing to greenhouse gas emissions, 'the impact ... depends on the amount and timing of the gases formed. Dams that flood large areas with large quantities of biomass (including underground biomass) generate GHG emissions' (Ishitani and Johansson, 1996, p 603).
6   As a result, mitigation costs involving hydropower options could well end up being higher than projected.
7   The 1996 study by Bacon et al (p 29), found that, in the 66 cases, the schedule slip was 28 per cent, which was 'very substantial'.
8   Carbon and particulate removal technologies for coal production and use, which would serve to lessen coal's harmful developmental consequences, could conceivably be developed at some point in the future. These technologies would, of course, reduce these numbers (and reduce the potential for mitigation as well).
9   The global warming potential of methane is 210 times that of carbon dioxide (UNEP, 1997, p 102).
10 Indeed, by virtue of the fact that we are capturing methane (a greenhouse gas) which would otherwise be released (in, for example, the baseline scenario), we have understated the global climate benefits of this action.
11 Within the SADC Energy Action Plan (SADC, 1997) there is specific mention of information and experience exchange. This will enable the sharing of best practice experiences in the sector, joint training activities, and through a joint regional power-sector investment drive, will significantly raise the level of trust among utilities. This, in practice, will increase the flow of regional consultations, which could lead to an even greater level of trust.
12 This could, of course, create concerns about the ceding of sovereignty among those in the host country!
13 Of course, some of this revenue might well flow out of these two countries since foreign entities (IPPs or some kind of governmental entities) might own the power facilities. Still, a good deal of it has the potential to remain within the countries concerned.

14 Even in the baseline scenario, all of these countries would earn revenue by 'wheel-ing' electricity. In the mitigation scenarios, however, the amounts would increase significantly.

15 In 1996, the Southern African Customs Union (South Africa, Botswana, Lesotho, Swaziland and Namibia) exported US$2543 million worth of goods to the other seven countries in the region, while importing only US$653 million worth from the same (IMF, 1997).

16 While Chapter 5 focused upon the differences in present day dollar values between each of our two mitigation scenarios, and our baseline scenario, we should still recognize the fact that the sheer quantity of resources required to implement any of these scenarios is immense. For example, the present value of the baseline scenario is close to US$100 billion (at 5 per cent discount rate), which does not even include replacement and refurbishment costs.

17 Of the 12 utilities in SADC (as of 1996), five are considered to have poor financial standing, four reasonable and only three good (SAD–ELEC and MEPC, 1996, p 76).

18 Indeed, this applies to much electricity investment: the SADC Energy Action Plan (SADC, 1997, p 19) reports that about 60 per cent of the envisaged regional investments (themselves valued at US$15 billion for the 1997–2005 period in generation, transmission and distribution) will need to come from private funds (equity and loans).

19 This is already happening in the region. The International Rivers Network has (as of January 1998) established campaigns in Namibia (the Epupa Hydroelectric Scheme) and Lesotho (the Lesotho Highlands Water project).

20 'Declaration of Curitiba: Affirming the Right to Life and Livelihood of People Affected by Dams', Curitiba, Brazil, 14 March 1997; http://www.irn.org/programs/curitiba.html

21 See, for example, World Economic Forum (1997), reporting on the meeting held in Harare. The 1998 meeting, meanwhile, was scheduled to take place in Windhoek, Namibia, from 17 to 19 May.

22 The 1997 Southern Africa Trade and Investment Summit was held in Gaborone, Botswana, on 18 and 19 November.

23 Reverend Leon Sullivan, chairman of the fourth African – African American Summit (Harare, July 1997), quoted in 'Comment' (1997).

24 An agreement to construct a hydropower facility in the DRC, for example, could require agreement among representatives from the country where the facility will be constructed (the DRC), the country through which the electricity will pass (the various wheeling states, perhaps Zambia, Zimbabwe and Botswana), the purchasing country (South Africa), and any other country that might have been involved in the financing.

Chapter 7

Other Regional Mitigation Options

Peter Zhou, Ian H Rowlands and John K Turkson

INTRODUCTION

As suggested by its title, this chapter sets out to explore a relatively broad agenda – namely, those regional options to mitigate global climate change that reside 'outside' of the power sector's supply side (the subject of the previous three chapters).[1] Though absolute gains in these non-power sectors may not be as large, and though data for undertaking analysis may not be as readily available, we should not be dissuaded from investigating their potential for climate change mitigation. Accordingly, we do just that.

The chapter is divided into two main sections. The first examines the transportation sector, exploring the extent to which regional cooperation in six different areas could potentially mitigate global climate change – namely, moving some cross-border transport from road to rail, electrifying regional railways, fuel substitution from gasoline to compressed natural gas, road paving and maintenance, building pipelines, and constructing bulk-supply depots. As far as possible, we estimate the mitigation potential and associated costs of each; broader developmental impacts and implementation prospects are also explored.

The second section of this chapter is much more qualitative in nature. Here we explore the potential for increasing the use of renewable energy and energy-efficiency technologies in the region. The discussion focuses upon a number of regional options that could achieve this goal – namely, bulk procurement, regional trade, harmonized standards and collaborative research and training. We do not undertake specific mitigation calculations. Instead, our purpose is to broaden the agenda and to explore alternatives that are rarely discussed in the context of climate change mitigation.

TRANSPORTATION IN SOUTHERN AFRICA

In southern Africa, many see transport as a key engine for regional development and integration.[2] Effective transportation links among the region's countries, businesses and peoples can support a range of economic, social and political ambitions for southern Africa.[3]

At the operational level, the region's transportation network is made up of road, rail, air and water (both ocean and inland) elements.[4] This section, however, will concentrate on road and rail transport since they are the major energy consumers in the region. Moreover, comprehensive data, which will allow for some kind of detailed analysis, exist for these modes of transport. Notwithstanding the fact that urban and national transport activities eventually contribute to regional transport services and energy demand, our analysis will be restricted to transboundary transportation involving two or more countries. To begin with, we will develop a broad baseline before we turn our attention to the six regional mitigation options.

Regional Transportation Baseline

In this study, we assume that future energy demand in the transportation sector will be directly related to per capita income in the region.[5] A snapshot of the situation in southern African countries today would suggest that, broadly, such a relationship exists (see Figure 7.1). Although not without its problems (we are only comparing development experiences across space, not across time as well), we nevertheless suggest that it is a reasonable way to generate a regional demand scenario to the year 2050 (following

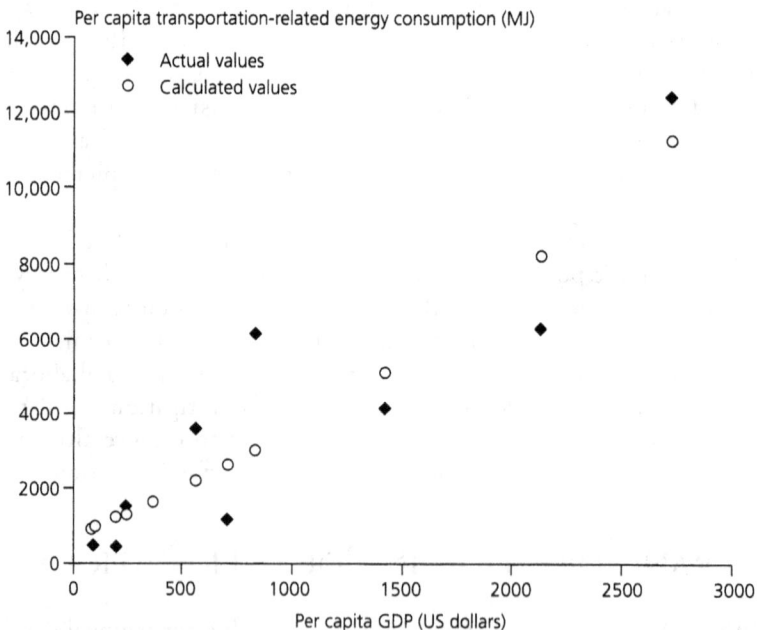

Sources: SADC, 1990; Directorate of Roads, 1995

Figure 7.1 *Per capita GDP and per capita transportation-related energy consumption for southern African countries, 1990, actual and calculated*

Table 7.1 *Actual and projected transportation-related energy demand levels for southern African countries, 1990, 2010, 2030 and 2050*

Country	Actual transportation-related energy demand, 1990 (TJ)	Projected transportation-related energy demand, 2010 (TJ)	Projected transportation-related energy demand, 2030(TJ)	Projected transportation-related energy demand, 2050(TJ)
Angola	11,960	67,594	135,708	227,537
Botswana	8,306	26,466	58,739	82,032
Lesotho	2,664	3,983	5,975	7,754
Malawi	4,013	18,334	28,935	40,982
Mozambique	7,332	30,200	56,185	83,632
Namibia	7,573	15,182	22,517	30,587
South Africa	472,004	830,187	1,480,279	2,069,979
Swaziland	4,926	4,757	7,786	10,487
Tanzania	14,032	46,502	70,970	95,586
Zambia	13,579	21,139	31,411	42,306
Zimbabwe	34,988	37,731	60,714	81,773
Regional total	581,377	1,102,075	1,959,220	2,772,654

Sources: SADC, 1990; Directorate of Roads, 1995; and authors' estimates

Michaelis, 1996, p 684). From this, we determine the relationship between per capita energy demand in transportation (E_t) and per capita GDP (GDP) to be as follows (at a 95 per cent confidence level): $E_t = 701.82 + 2.28°GDP + 0.000609°GDP^2$. Figure 7.1 also plots the calculated energy demand levels, using this relationship.

To generate an energy demand profile for the next half century, two regional projections are required: GDP and population. For the former, we make estimates following some of the work of the World Bank. Regionally, we expect annual growth in GDP to be approximately 3.2 per cent to the year 2010, 2.6 per cent to 2030, and 1.7 per cent to 2050.[6] For the latter, meanwhile, we use projections from national statistical offices, supplemented by our own estimates. Annual population increase for the region is thus projected to be 2.7 per cent to 2010, 2.1 per cent to 2030 and 1.4 per cent to 2050. Together, these figures allow us to calculate national transport energy-demand figures (see Table 7.1).[7] Table 7.1 clearly shows how South Africa is expected to dominate the demand for energy in this sector: the country accounted for 81.2 per cent of total regional demand in 1990, and the corresponding figure for 2050 is projected to be 74.7 per cent.

Given that almost 95 per cent of the energy powering the transportation sector is derived from petroleum (Zhou, 1997), we can calculate carbon dioxide emissions by noting that gasoline, diesel and other petroleum products have emission factors of the order of 70 to 75 kilogrammes of carbon dioxide per GJ (UNEP, 1997, p 102). Consequently, for the region

as a whole, carbon dioxide emissions from the transport sector are expected
to be 80 million tonnes in 2010, 142 million tonnes in 2030 and 201
million tonnes in 2050.[8] For the sake of comparison, recall (from Chapter
5) that our baseline scenario in the power sector projected that 459 million
tonnes of carbon dioxide would be generated by the region's power stations
in the year 2050. Consequently, though the transportation sector appears to
be set to emit less than half as much greenhouse gases in the future, the
numbers involved are still significant: 201 million tonnes of carbon dioxide
is almost double that emitted by the United Kingdom's transportation
sector in 1990 (121 million tonnes carbon dioxide; UK National
Communication, 1994).

Regional Transportation Mitigation Options

In this section, we examine six possible regional mitigation options for
southern Africa. These may help to illuminate the wider potential for greenhouse-gas emission reductions in the transportation sector as a whole (and
particularly those opportunities that exist at the regional level).[9] These
options – which are presented and considered independently of each other –
are as follows:

- moving from road to rail;
- electric trains replacing diesel trains;
- fuel substitution from gasoline to compressed natural gas;
- paving and maintaining the regional road network;
- building pipelines;
- constructing bulk supply depots.

For each possible mitigation option, we examine its baseline character,
identify and analyse the modifications that define the particular mitigation
scenario and highlight some potential developmental consequences and
implementation challenges. Where possible, estimates of greenhouse-gas
reduction levels and costs are also made; where data limitations preclude
this, a more general discussion of the option is undertaken.

Moving from Road to Rail

Much of southern Africa's transportation is concentrated along a number of
'corridors', each of which links part of the region's hinterland with at least
one of its seaports. Six southern African countries are landlocked and hence
need effective links of rail and road to and from these seaports (all the more
important given the amount of trade that these countries carry out with non-
African countries). Listed in order of decreasing traffic levels, the six major
corridors in the region (and the respective seaport(s) for each) are as follows:

- Southern Corridor (South African seaports, collectively known as the *portnet*);
- Eastern II Corridor (Beira and Maputo in Mozambique);
- Northern Corridor (Dar es Salaam, Tanzania);
- Eastern I Corridor (Nacala, Mozambique);
- Western I Corridor (Walvis Bay, Namibia);
- Western II Corridor (Angolan seaports).

These corridors offer integrated transport systems using rail and road to transport both dry and liquid cargo. In a few cases, pipelines are used to transport petroleum products as well.

South African seaports (and hence the Southern Corridor) dominate the region. In 1993, they accounted for 89 per cent of the region's loaded and unloaded goods, while seaports in all of the other countries combined handled the remaining 11 per cent (SADC–SATCC, 1996). This figure is all the more remarkable in light of SADCC's conscious effort to use other corridors during the apartheid era. Indeed, given South Africa's growing links with its regional neighbours, an even higher proportion of the region's goods are probably being channelled through this corridor today. Consequently, irrespective of any potential climate change demands and policies, regional transport planners will have to consider how the region's other corridors might be improved, in order to relieve the pressure on the Southern Corridor.

Let us turn briefly to the region's other corridors. Funds to rehabilitate Mozambican ports were given priority by the SADC(C) countries before South Africa's independence. Hence, landlocked countries in the region strove to channel many of their goods through Mozambique – mainly through the port of Beira – in an effort to isolate South Africa. Notwithstanding this effort, however, Mozambican ports handled only 2 per cent of the region's goods in 1993 and 1994 (SADC–SATCC, 1996). The Northern Corridor, meanwhile, is used by Zambia, Malawi and Tanzania (with the bulk of goods at Dar es Salaam originating in, or being destined for, Tanzania itself) (SADC–SATCC, 1996). The corridors in the western part of the region, finally, are the least developed and are mainly used by businesses within Namibia and Angola. (Angola's ports are, naturally, dominated by petroleum products.)

As with many issues in the region, transportation is clearly dominated by South Africa. Moreover, not only is South Africa dominant with respect to seaport activity (as noted above), but many of the goods that pass through South African seaports do not end up moving regionally – the only southern African country in which they travel is South Africa.[10] Thus, although many of the region's corridors are international in nature, much of the actual activity is national (at least in terms of the southern African context). As a result, it is clear that considerable potential for emission reductions in the transportation sector may exist at the national level, within

South Africa itself. It is worth reiterating, therefore, that regional options should not be examined at the expense of parallel national options: the latter may hold the greater potential.

Nevertheless, our task in this book is to consider possible regional options. Given that there is some international activity in many of these corridors, worthwhile regional mitigation options could conceivably exist. To begin our exploration, we propose a specific one – namely, the movement of goods from roads to rails within these international corridors. To investigate this more closely, we examine one part of the Eastern II Corridor: the Beira route.

Between 60 and 70 per cent of goods travelling through the Mozambican port of Beira have either come from, or are heading to, Zimbabwe. The length of this route, from Beira (in Mozambique) to Mutare (in Zimbabwe), is approximately 345 kilometres. The mass of dry cargo carried on this route during recent years has been as follows: 1400 kilotonnes (in 1992), 1562 kilotonnes (in 1993) and 1290 kilotonnes (in 1994). Although fluctuations can be partially explained by the presence or absence of drought in the region – which in turn influences the amount of drought relief food required, as well as the quantity of agricultural goods exported – present regional trends in trade, as well as developments themselves within the corridor, suggest that traffic levels will grow in the future.[11]

It is, of course, difficult to project future traffic levels. However, we assume that they will be proportional to Zimbabwe's per capita GDP levels.[12] Taking both an average cargo-to-population ratio, and an average cargo-to-GDP ratio, for the years 1992, 1993 and 1994, we use this to estimate future dry cargo masses. In our baseline, we will assume that 80 per cent of this is carried by rail and 20 per cent by road, throughout the study period. The split in 1994 was 74:26. These projections are presented in Figure 7.2.

Our mitigation scenario consists of developing sufficient rail capacity so that 100 per cent of the dry cargo can be transported by rail. Consequently, the differences between our baseline scenario and our mitigation scenario will be the differences between using road equipment (trucks and trailers) and rail equipment (diesel engines and wagons) to carry 20 per cent of the corridor's dry cargo. To calculate this difference, firstly, in terms of reduced carbon dioxide emissions, note that the energy intensity of fully laden diesel trains averages 0.23 megajoules per tonne-kilometre (MJ/t-km), while the comparable figure for truck trailers is 0.85 MJ per tonne-kilometre (ETSU, 1995); moreover, note that diesel (which we assume will power both trains and trucks) has a carbon dioxide content of 74.1 kilogrammes per GJ (UNEP, 1997, p 102).

We make several other further assumptions. In the baseline scenario, for instance, each truck is able to carry a cargo weighing 34 tonnes, and the truck itself weighs an additional 20 tonnes; moreover, a truck is able to make the return journey between Mutare and Beira in five days. For the

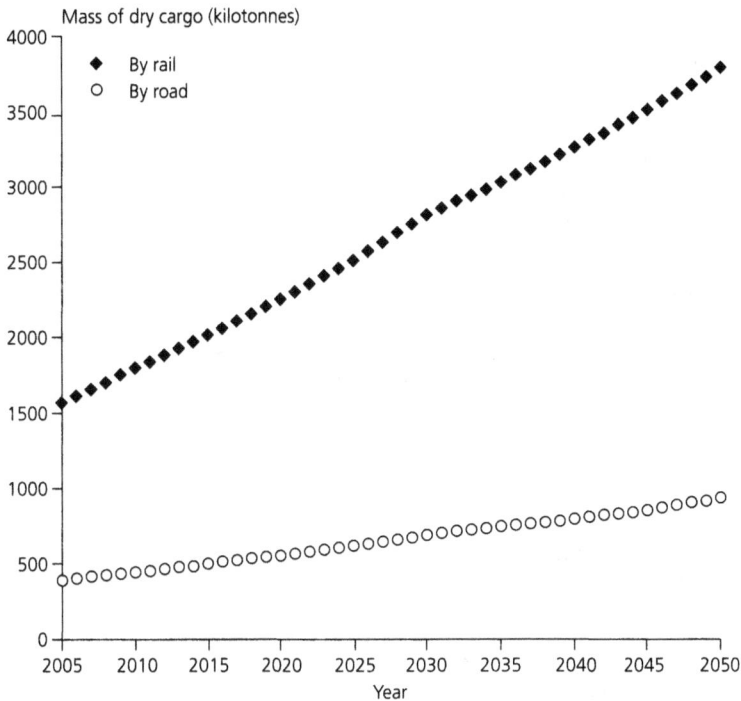

Figure 7.2 *Projected movement of dry cargoes in the Beira Corridor,*
by rail and road, 2005–2050

mitigation scenario, meanwhile, we assume that each train can carry 600 tonnes of cargo, and that the train's own weight is an additional 696 tonnes.[13] It, however, takes ten days to complete the return journey.[14] Finally, for both scenarios, we assume that all rail and road journeys are made with vehicles that are fully loaded (for both the eastbound and westbound journeys), and that all cargo travels the entire length of this route.[15]

Given these assumptions, we are able to estimate reductions in emission levels. If the project is operational in 2005, we find that approximately 665,000 tonnes of carbon dioxide are abated during the subsequent 46 years of operation (including approximately 21,000 tonnes during the final year of our study, 2050). This final year figure is equal to just over 0.01 per cent of all transportation-related emissions in the region for that year. While this suggests that the reduction is extremely small (as would a comparison with Chapter 5's study of the power sector, which involved annual savings of many million tonnes), it would still have some degree of beneficial impact upon the global climate.

Turning to cost, the differences between the baseline scenario and the mitigation scenario relate to the need to operate rail instead of road. In order to obtain a general estimate of cost, we will consider the key differences involved in moving from the baseline scenario to the mitigation scenario – namely, the need to buy additional trains (including ten in the year 2005; each of these, as well as each of the subsequent ones required, we will cost at US$3.5 million). Nevertheless, there is no need to buy additional trucks (which we will cost at US$150,000 each). Differences in the cost of the diesel to move these modes of transport will also be considered.[16] Among the many cost elements we are not considering are any differences in annual operation and maintenance expenses (apart from fuel prices);[17] the revenue that might be received from the sale of the disused trucks in the year 2005 (all 88 of them, following our assumptions), and the implications of different life spans of the two modes of transport.[18]

With these many provisos, we nevertheless find that the difference in net present value is approximately US$5.7 million: while the capital costs of the new trains are much greater than those of the new trucks, the lower fuel costs required for the trains make the final difference relatively small. Indeed, this suggests that, at a 5 per cent discount rate, carbon dioxide emission reductions can be realized at a cost of US$8.58 per tonne in 1997 prices.[19]

In addition to aiding the global climate, removing traffic from the region's roads could have a number of local benefits. For one, the level of local air pollution would fall, which is particularly significant given that the transport sector is known to generate roughly half of all urban air pollutants (Sperling and Deluchi, 1993). Inferior air quality can be a significant contributor to high mortality rates (WEC, 1995a) and incidents of illness, particularly in urban areas, when emission levels exceed WHO recommended limits.

Less traffic on the roads could also lead to fewer accidents. Though we do not have figures for the Beira route, figures for Zimbabwe reveal that there were 1066 deaths as a result of traffic accidents in 1992 (Heggie, 1995). Traffic congestion could also be eased: such tie-ups result in unnecessary energy consumption, increased pollution and delays to work; each has its own detrimental impact upon development and development prospects in southern Africa. Finally, pursuit of this mitigation option would permit significant savings of foreign exchange, which would otherwise have been spent on importing diesel from outside of the region. Quantifying benefits like these would reduce the cost of many of these transportation mitigation options.

Implementing this proposal – indeed, implementing any regional transportation-related activity – could pose particular challenges. Although there is a well-established regional transportation entity – namely, the Southern Africa Transport and Communications Commission (SATCC), located in Maputo, Mozambique[20] – it can only play a coordinating role since national governmental institutions continue to take the key decisions.[21] Consequently, the usual sets of challenges associated with regional policies – as identified in Chapter 2, could arise here.

On the other hand, however, it might be much easier to implement specific project proposals in transportation – such as the purchase of additional rail capacity – than similar projects in electricity generation. Why? Because our mitigation proposal does not represent a dramatic departure from the baseline, or the status quo. Granted, those who own or operate road haulage companies may well feel aggrieved, but the mitigation proposal is basically a scheme to make existing corridors 'better'. If we were to consider moving traffic from one corridor to another, which would more directly damage particular national interests (for example, governments would potentially lose toll revenues and businesses within the corridors would lose customers) then stronger opponents may become more vocal. However, given the traditional and ongoing support, both inside and outside the region, for corridor development (for example, World Bank, 1995), challenges to implementation would probably be muted. Indeed, this is the kind of project with which SADC(C) has traditionally been most successful.

The Beira route is, of course, not the only candidate for this particular regional mitigation option. In theory, similar calculations could be made for the other international corridors in the region that have both rail and road in parallel. However, reliable data for each (particularly with respect to how much of the cargo travels regionally and the quantities of freight being moved by rail and road) are not always readily available. Nevertheless, casual observations suggest that the Tazara Corridor (which is part of the Northern Corridor, between Zambia and Dar es Salaam)[22] may offer considerable potential.[23] It may, however, be the case (as we note above) that few of the goods actually travel outside of Tanzania into any other part of the region.

Electrifying Rail in the Corridors

A second option involves changing the rail systems – more specifically, moving from diesel-powered to electric-powered trains. This would, of course, only be a climate change mitigation option should the electricity have been generated by some kind of carbon-free process (perhaps hydropower, perhaps renewable energy). Here, we will assume that the electricity comes from hydropower facilities in the region that are dedicated to this purpose. To explore the potential for this option, we begin by considering the Beira route in the Eastern II Corridor. Our mitigation option consists of replacing all the diesel trains that are required to carry rail cargoes in our baseline (see previous section) with electric trains.

We first consider reducing emissions. Given that the average energy intensities for fully laden electric and diesel trains are 0.27 MJ/t-km and 0.23 MJ/t-km respectively (ETSU, 1995), there is no actual energy saving as such. The benefit, however, derives from the fact that we are replacing carbon-intensive diesel with carbon-free electricity. Consequently, by knowing the carbon intensity of diesel, calculating the quantity of carbon dioxide abated is relatively straightforward. The total for the period (again assuming implementation in 2005, along with the other assumptions outlined above) is approximately 1.5 million tonnes (including approximately 48,000 tonnes in 2050).

With respect to cost, there are both fixed costs and variable costs to consider. The most significant fixed cost would be the electrification of the railway line, which we estimate to be US$500,000 per kilometre (Tractebel-Belgium, personal communication). Given that the length of line is 345 kilometres, the required investment comes to US$172.5 million. Another fixed cost would be constructing the power station to provide the electricity. Given that approximately 212 million kilowatt-hours (kWh) of electricity are required during the final year of our study period (2050), we assume that a 41 megawatt (MW) mini-hydropower facility would be able to meet the demand throughout the 46-year period (working at a 60 per cent load in the final year, and at a lower load level in previous years). Including requisite transmission lines, we estimate the cost of such a facility to be US$82 million (equivalent to US$2000 per installed kW). Finally, purchase of a fleet of new electric trains in the year 2005 has to be included: 29 trains at US$3.5 million each adds US$101.5 million to the bill.[24] Therefore, total fixed costs amount to US$356 million, and we assume that these costs will be incurred in the year 2005.

For variable costs, meanwhile, we assume that the operation and maintenance of the track would be the same in each case (that is, for either diesel-powered or electric-powered trains). Differences arise by virtue of the different fuel being used in each case. In the case of the baseline scenario, this is the cost of diesel (the price of which we have already noted). In the case of the mitigation scenario, meanwhile, it is the cost of maintaining the hydropower station, which has been constructed to supply electricity to the railway. We estimate this to be US$820,000 annually (based on Chapter 5, this is equal to 1 per cent of the original capital costs of construction). Calculations reveal that, although annual variable costs for the mitigation option are much lower, the initial capital costs are still large enough to negate these savings. As a result, the difference in the net present value between the two scenarios is (in 1997 values, at a 5 per cent discount rate) approximately US$59.2 million. Given the carbon dioxide savings noted above, the corresponding cost per tonne of carbon dioxide abated is US$38.28.

We could, in addition, consider moving the remaining dry cargo (20 per cent of the total mass) off of the roads and on to the electrified rails. This would increase carbon dioxide emission savings by an additional 1.05 million tonnes (including approximately 33,000 tonnes in the final year of our study). Additional costs, however, would be incurred by virtue of the larger hydropower facility required (51 MW, instead of 41 MW) and the additional capital costs (more trains); but savings would be accrued by virtue of being released from the need to purchase diesel, and savings from the additional trucks not purchased. Calculations reveal that the net present value is US$71.7 million, so unit costs work out to US$27.57 per tonne of carbon dioxide abated (which is expected, given that the most expensive undertaking – that is, the construction of the electrified line – has already been completed).

Electrifying the corridor's railways would generate a number of additional local benefits – of the kind discussed in the previous section. Indeed, they could well be larger in magnitude, given that we are not only reducing air pollution from road traffic, but from rail traffic as well. However, the problems associated with hydropower facilities (see Chapter 6) could serve to dampen some of the local enthusiasm for this proposal.[25]

Fuel Substitution from Gasoline to Compressed Natural Gas

As outlined in Chapter 4, southern Africa has considerable reserves of natural gas. This resource could be exploited as a mitigation option in the transportation sector: substitution of compressed natural gas (CNG) for gasoline in vehicles would serve to reduce emissions by virtue of the lower carbon intensity of natural gas (56.1 kilogrammes of carbon dioxide per GJ, compared with 69.3 kilogrammes of carbon dioxide per GJ for gasoline, a difference of 24 per cent; UNEP, 1997, p 102).[26] An option that we will consider here, therefore, is that 1 per cent of the cars and light duty vehicles in South Africa are, in the year 2005, converted to use CNG.[27]

Our calculations are quite straightforward. From previous work in this chapter, we know the energy consumed by the transportation sector in South Africa (or, at least our projections of the same). We will assume that 62 per cent is attributable to cars and light duty vehicles.[28] We propose to convert 1 per cent of these, each of which will result in a decrease in the level of carbon dioxide emissions by 20 per cent. Therefore, a broad estimate is simply a reduction in South Africa's total emissions by the product of 62 per cent, 1 per cent and 20 per cent – namely, 0.124 per cent. Calculations reveal that savings for the period of 2005 to 2050 amount to just under 5.4 million tonnes of carbon dioxide. With emissions reductions of approximately 178,000 tonnes in the year 2050, this is equivalent to almost 0.1 per cent of total transportation-related regional emissions in that year.[29]

Cost is somewhat more difficult to calculate. We consider, first, the fixed costs. The IPCC study (Michaelis, 1996, p 698) reports that the cost of modifying a gasoline car to use CNG was, in the case of New Zealand in 1984, approximately US$750. Given that technological developments may have dampened any rise in cost owing to inflation, we take this as our price per vehicle today; IPCC figures support this (Michaelis, 1996, p 696).[30] An estimate of the number of cars that will need to be modified in the year 2005, however, is necessary before we can calculate these fixed costs. To derive this, we estimate that 80 per cent of South Africa's road transport fleet is made up of either cars or light duty vehicles. (This is higher than the 62 per cent share of energy and emissions that we estimated, because heavy trucks will be more intensively used than the lighter duty vehicles in which we are interested.) Given that there were 5,030,743 vehicles in South Africa in 1991 (Heggie, 1995, p 153), we calculate that there will be almost 6.4 million such vehicles in 2005 (assuming growth rates similar to South

African transportation-related energy consumption as a whole). Consequently, just under 64,000 vehicles will require modification, which will, in turn, necessitate an investment of almost US$48 million.[31]

Variable costs, however, will allow some of these initial capital costs to be recouped. Here, and following the IPCC (Michaelis, 1996, p 696), we take the key variable cost to be the difference in the expense of CNG and gasoline, which is about 20 per cent (or US$0.05 per litre gasoline equivalent). Using projected gasoline prices, we calculate annual savings to be US$3 million in 2005 and to rise to US$10.5 million in 2050.[32] Nevertheless, the initial capital outlay is so immense that the difference in the net present value of the two scenarios (at 5 per cent discount rate, and at 1997 prices) is US$7.4 million. Unit costs are US$1.37 per tonne of carbon dioxide abated.

A broader developmental consequence of this scenario is that more resources will again be kept inside of the region: instead of paying extra-regional actors to supply petroleum products, regional players could be reimbursed for the provision of natural gas.[33] Moreover, because the CNG could also be piped for other purposes – for example, electricity generation and chemical industrial uses – the development of regional CNG resources and infrastructure could advance development in southern Africa more broadly.

Implementation challenges, however, abound (though they are not necessarily uniquely regional). In the New Zealand experience that the IPCC recounts (Michaelis, 1996, p 698), take-up of the option was, at times, quite low. Generally, drivers may well fear that CNG will not be as widely available as promised. If availability does not materialize, then they will be left with a means of transportation with a relatively restricted mobility range! Strategies for ensuring adequate supply would need to be seriously examined.

Paving and Maintaining the Regional Road Network

When an unpaved road is paved, energy savings of as much as 50 per cent can result (BTP, 1983). Similarly, when a poorly maintained road is repaired, less energy is consumed to travel the same distance. Because a lower level of energy consumption also reduces the levels of carbon dioxide emissions, we consider the paving of roads – and the proper maintenance of roads more generally – to be a climate change mitigation option. Furthermore, we present it as a regional option because we envisage that roads in the region's corridors would receive priority (as is, indeed, the case with the current corridor rehabilitation programmes, managed by SADC through the SATCC).

Given the relative lack of data on vehicle traffic in these corridors, we do not make any explicit effort to calculate emission reduction potential, and the cost of the same. Instead, we simply present some data about the state of the region's roads in Table 7.2 to suggest that there is considerable potential.[34]

The World Bank has estimated that regular maintenance of roads makes economic sense, irrespective of climate change benefits: 'When a road is not

Table 7.2 *Length and condition of countries' main road network, 1989*

Country	Total length (km)	Paved length (km)	Percentage of total length that is paved	Percentage paved length in 'good' condition (estimated)
Angola	15,811	7,942	50	na
Botswana	17,867	2,831	16	94
Lesotho	2,346	600	26	53
Malawi	9,963	2,520	25	56
Mozambique	13,308	4,600	35	19
Namibia	na	na	na	na
South Africa	62,053	57,034	92	na
Swaziland	2,757	689	25	35
Tanzania	28,011	3,349	12	39
Zambia	20,783	6,396	31	40
Zimbabwe	18,434	8,261	45	70

Source: Heggie, 1995

maintained – and is allowed to deteriorate from good to poor condition – each dollar *saved* on road maintenance *increases* vehicle operating costs by $2 to $3. Far from saving money, cutting back on road maintenance *increases* the cost of road transport and raises the net costs to the economy as a whole' (Heggie, 1995, p 1, examining sub-Saharan Africa as a whole). Moreover, national benefits for developing different parts of the region, thereby increasing equity ambitions, might prove to be other attractions for southern Africa.[35] However, when judged solely in terms of costs per unit of carbon dioxide abated – in light of estimated paving costs of US$70,000 per kilometre (Government of Botswana, Ministry of Roads, personal communication, 1997) – the expense might appear prohibitive.

Building Pipelines

Petroleum products are used throughout southern Africa. In many areas (including those with the highest demand), they are provided by pipelines. For example, in the case of the Gauteng region (South Africa's industrial heartland), refined petroleum products, crude oil and gas are piped from Durban; while in the case of Harare, refined petroleum products are piped from the port at Beira; finally, Zambia receives crude oil by pipeline from Tanzania. However, other land-locked countries in southern Africa – namely, Botswana, Lesotho, Malawi and Swaziland – rely on road and rail transport systems to receive their petroleum products from elsewhere in the region.

Increased pipeline use could serve as a regional mitigation option since this would eliminate the need for petroleum products to be transported by either road or rail. To explore further the potential of this mitigation option, let us examine the Pretoria (South Africa)–Gaborone (Botswana) route.

Today, Gaborone is supplied with refined petroleum products from the Gauteng area (which is not only at the end of the Durban pipeline, but is itself also home to refineries). These products travel between the two countries by either diesel trains (a distance of 600 kilometres) or road freight (400 kilometres). The split between the two modes of transportation is approximately 60:40. Our mitigation option consists of constructing, in the year 2005, a 400-kilometre pipeline between these two locations; the pipeline subsequently carries all of Botswana's petroleum products.

In order to calculate the reduction in carbon dioxide emissions, we draw upon some data already presented in this chapter – namely, the demand for petroleum products in Botswana (which was based on GDP and population growth rates), the energy intensity of diesel trains and truck trailers (which we took to be 0.23 and 0.85 MJ/t-km respectively), the mass of a truck and its cargo (20 tonnes and 34 tonnes respectively), and the mass of a train and its cargo (696 tonnes and 600 tonnes respectively). We add to this the additional assumption that trucks can complete the return trip between Pretoria and Gaborone in five days, and trains in ten days. For the sake of our mitigation option, we also assume that the pipeline operates ideally – in particular, that it does not leak at all. Moreover, we assume that the electricity to power the pipeline's pumps would be generated by some sort of carbon-free process (either mini-hydro, or renewable energy of some kind). Granted, these conditions are very favourable for the mitigation option, but they will nevertheless provide us with an order of magnitude estimate for the carbon dioxide reductions.

Calculations reveal that, during the 46-year period under study, the quantity of carbon dioxide abated amounts to almost 2.6 million tonnes (including almost 90,000 tonnes in the year 2050). Costs are more difficult to calculate, however. After making a number of assumptions, we nevertheless still leave out the costs of annual operation and maintenance of elements in both scenarios – that is, keeping the trucks on the road and the trains on the rails in the case of the baseline scenario (including the cost of fuel), and maintaining the integrity of the pipeline in the case of the mitigation scenario (including the cost of the hydroelectricity).[36] Consequently, we simply compare the net present value of differences between the capital costs of the two scenarios. Again, our motivation is to provide some initial order of magnitude estimate. We find that this amounts to almost US$164 million, in terms of net present value at a 5 per cent discount rate. Consequently, our reductions are realized at an estimated cost of US$63.41 per tonne of carbon dioxide abated.

As was the case in some of our previous discussions, there could well be other benefits arising from successfully implementing this option: better local air quality and less deterioration of the region's roads (as a consequence of reducing road traffic) and improved balance of payments (as a consequence of reducing the need for imported fuel) are two of the most significant. Given that most of the region's demand centres are already supplied by pipeline, however, the potential for replicating these savings elsewhere is relatively small. Nevertheless, a pipeline from Mozambique to Malawi may be one candidate for further investigation.[37]

Constructing Bulk Supply Depots

Closely linked with the pipeline as a potential mitigation option is the strategic location of bulk supply depots (BSDs) to hold petroleum products. The underlying rationale for this strategy is, once again, to reduce the distances over which petroleum products are transported – from ports or refineries to consuming centres.

Within southern Africa, the existing pattern of transporting petroleum products is from south to north (that is, from South Africa to Botswana, Zimbabwe and Zambia), south to west (that is, South Africa to Namibia) and from east to west (that is, from Mozambique to Zimbabwe and, to a limited extent, Malawi, and from Tanzania to Zambia). Though some petroleum is transported by pipeline (noted above), much is either by rail or road. As an alternative to constructing new pipelines (again, see above), a regional strategy of siting BSDs at different locations – selected to minimize distances over which petroleum products are transported either by rail or road (or both) – has the potential of reducing greenhouse gas emissions from the transport sector.

Consider, once again, the case of the Pretoria–Gaborone route. Botswana's petroleum product requirements are presently supplied from a BSD in Pretoria, South Africa. Though the Pretoria depot itself is supplied through a pipeline from Durban, the onward journey to points in Botswana is undertaken by either road or rail. Were a BSD of some sort to be constructed nearer to Gaborone, products could be sent there more efficiently and subsequently distributed from there.

In late 1996, it was reported that southern Africa would soon experience shortages of gasoline and, by 1998 or 1999, diesel as well (Paxton, 1997). This has resulted primarily from an increase in the consumption of petroleum products in South Africa in recent years. Nevertheless, there is much debate about whether new refineries should be constructed in the region – and if so, where and when. While the potential for increased capacity exists in South Africa, some expect local environmental groups to resist successfully any such expansions. As a result, other areas, such as Angola and Mozambique, are being considered as potential locations for future refineries. Irrespective of the relative merits of different locations, attention upon the market for petroleum products means that discussion about the location of BSDs is a timely one. Considering the two together could prove to be strategic: the future distribution of refineries and BSDs in the region could also be presented as an integrated mitigation option for the region.

RENEWABLE ENERGY AND ENERGY-EFFICIENCY TECHNOLOGIES

In this section, we turn our attention to a specific goal: namely, increasing the uptake and use of renewable energy and energy-efficiency technologies in the region. We do not investigate the particular characteristics of the technologies we are considering since our aim is to launch a general discussion. Nevertheless, included in what we are calling renewable energy technologies would be bio-, solar and wind energy (Karekezi and Ranja, 1997), while included in energy-efficiency technologies would be products such as compact-fluorescent light bulbs and CNG-powered vehicles. Given that we are not investigating specific technologies, our discussion is entirely qualitative, striving merely to identify alternatives rather than to cost rigorously the same. We nevertheless believe that this is an important exercise: discussion about such a pan-regional strategy is at a relatively early stage. In the rest of this chapter, we examine four regional methods for promoting the increased uptake and use of renewable energy and energy-efficiency technologies. These are: bulk procurement, regional trade, harmonized standards and collaborative research and training.

Bulk Procurement

We propose here that bulk procurement − coordinated amongst entities in two or more countries − could serve as a mitigation option. For example, governments could purchase items that mitigate global climate change, thereby accruing economies of scale on the purchase side (lower unit costs). This would allow them to pursue a mitigation option more cheaply than if they had done the same unilaterally. Moreover, once contracts to provide such bulk supplies were offered, a range of potential suppliers could be encouraged to compete for selection (see, for example, Mullins, 1996). With such a carrot on offer, technological advances and lower costs could well be forthcoming. Given the volume of purchasing executed by governments, the quantities involved could be considerable.

Regional Trade

Constructing an international free trade area could also reduce the cost of either a renewable energy technology or an energy-efficiency technology. By creating a free trade area, not only might production inputs be cheaper (because regional suppliers can supply goods at lower costs), but economies of scale in terms of production runs could be captured because of the increased size of the regional market.[38] In the end, therefore, an existing producer may well be able to supply the technology more cheaply, or new producers may be encouraged to emerge. The point is simply that coordinated regional action – in this case, the construction of a free trade area – may cause the net benefit or cost of that particular mitigation option to shift, compared with the result of action solely at the national level.

The construction of a regional free trade area is clearly already a part of the southern African agenda, though primarily for other non-climate reasons (see Chapter 3 of this book). However, what this suggests to us is that this mitigation option fits with the region's own development aspirations. Moreover, the importance of regional trade in climate-relevant goods has already been noted in the SADC Energy Action Plan (from 1997). Among the prescriptions included in that document are:

- 'facilitating the trade of products and services in the region and internationally by initiating promotion for sufficient standardization and the elimination of trade barriers, and following up the training activity supporting the operationalisation process' (SADC, 1997, pp 48–49); and
- 'expanded markets for energy efficiency and conservation' (SADC, 1997, p 53).[39]

Again, this provides evidence that development of more open markets in southern Africa is already on the regional agenda.

For the sake of the global climate, benefits would arise from the increased use of the renewable source of energy, or the increased use of the technology for energy efficiency. Though the net costs of such an option could well, in the long run, be negative, there could be some kind of start-up costs associated with it: perhaps the tangible costs of supporting organizational development of the free-trade area (for example, the establishment of a secretariat or the training of customs officers), as well as the side-payments that might be necessary to keep national treasuries happy in the short term (perhaps financial compensation for lost revenue from tariffs). Though we do not quantify these respective costs and benefits here, efforts to do so could conceivably begin.

Of course, the implementation issues associated with such an option are plentiful. Indeed, given that much of the study and practice of regional cooperation has revolved around efforts to promote economic integration, most of our discussions in Chapters 2 and 3 are relevant here. Moreover, the fact that this mitigation ambition (increased uptake of new technologies) is delivered by means of a policy, it may prove challenging to attract support in the climate change context. To date, such policies have rarely been raised in the climate change debate, primarily because it is difficult to quantify associated costs and the amount of carbon dioxide abated. It is one thing to price a pipeline (an example of a particular project), but it is altogether another thing to cost a free trade zone (an example of a broader policy). Although economists do attempt to do so, debates about methodology and operationalization still abound. This, however, should not preclude further consideration.

Harmonized Standards

As part of a regional trade agreement – though not necessarily contingent upon it – countries could work together to introduce common standards of some kind. An Expert Group of Annex I countries investigated a range of Policies and Measures for Common Action, among them the prospect of common standards in energy-intensive goods (Mullins, 1996). Their motivation was manifold.

For one, the harmonization of test protocols could reduce the cost to manufacturers of making their products ready for the market: by releasing them from the need to undertake multiple tests to meet varied standards in multiple markets, only the costs of reaching one set of standards would have to be met. This would open markets since such intraregional non-tariff barriers (certification, for example) would be eliminated. Indeed, harmonization would lessen trade distortions more widely. With a more open regional market, larger production runs could be possible, which in turn would encourage the wider availability of cheaper and more efficient products (Mullins, 1996, pp 29 and 6).

However, while this may be fine for Annex I (developed) countries, will it be as relevant for a group of non-Annex I (developing) countries? Indeed, given the dominance of South Africa in the southern African economy, as well as the relatively low level of intraregional trade, harmonization of standards across southern Africa may have relatively little impact. The consequences of such a policy are, at this point, difficult to predict. However, what is evident is that, firstly, many of the region's economies are growing at impressive rates and, secondly, the level of intraregional trade is also growing considerably. Consequently, the potential for future mitigation may be considerable (and remember that we are primarily interested in the future level of net greenhouse gas emissions).[40]

Moreover, there could well be secondary developmental benefits arising from such an action. For one, energy efficiency could reduce demand for coal-generated electricity – and accordingly accrue the local advantages associated with that action (see Chapter 6). Harmonization could also serve to lend support to the broader regional project being undertaken by SADC, for the same reasons as outlined under our discussion of regional trade arrangements above. Indeed, additional benefits of harmonization suggest that it may simply be good for business, irrespective of global climate policies and demands. Among the benefits of such action are:

> ...*clarifying production and marketing requirements in all member economies against recognised and agreed benchmarks, increasing certainty*

among market suppliers in terms of production planning, agreeing test proto-
cols, with the potential for reduced testing and retesting requirements (and
hence costs) in multiple; and increasing certainty among regulators on accred-
itation procedures and quality assurance process, further reducing technical
and administrative requirements and costs. (Action Program for
Energy of the Asia Pacific Economic Cooperation, APEC,
quoted in Mullins, 1996, p 29)

As with many regional aims, however, there may be considerable implementation challenges. Indeed, virtually all of the implementation barriers already
identified – both in Chapter 2, and above in this chapter – will also present
themselves in this case. Nevertheless, as with many of the options being
presented here, it may warrant further investigation.

Collaborative Research and Training

Regional cooperation in the area of innovation is another potential means of
implementing this regional mitigation option. This would allow different
entities – probably, in this case, firms – to share risk, reduce costs and
increase the effectiveness of initiatives. It has been argued that 'common
actions in this area might further help to share the costs, risks and benefits
associated with experimentation, and provide opportunities for additional
exchanges of ideas and experience which would facilitate governments and
others in identifying and evaluating key opportunities' (Annex I Expert
Group, 1996). Basically, the exercise of putting more heads together might
identify better mitigation options for southern Africa.

Indeed, information and experience exchange is a priority that the southern African region has already set for itself, irrespective of climate change
developments: in the 1997 SADC Energy Action Plan, this was the identified as the 'first priority area ... [to] concentrate on ...' (SADC, 1997, p
2). As a result, it is clear that such collaboration fits with regional development goals. It could, conceivably, also advance climate change mitigation
ambitions, though it is certainly premature to quantify such benefits.[41]
Again, we flag it in an effort to develop a full agenda for further consideration – the examples could be anything from convening a seminar to sharing
experiences on the development of national markets for renewable energy
technologies to establishing a regional training centre for national customs
officers (to facilitate the efficient movement of goods within southern
Africa). The possibilities are virtually endless.

SUMMARY

This chapter has identified a number of regional options for global climate change mitigation that do not relate to the supply of electricity. Many of the options involve transportation – not surprising, given both this sector's greenhouse gas emissions and its regional nature. Within this discussion (undertaken in the first main section of this chapter), we examined a number of specific options; where possible, we also quantified the mitigation potential as well as the cost of the same (at least to gain some sort of order of magnitude estimate). These options ranged from the relatively narrow and specific (for example, increasing the use of railways along a particular regional route) to the broad and general (for example, substituting CNG for gasoline in some of South Africa's vehicles). We did not set out to be exhaustive; nor have we been, by any means. Indeed, not only are there additional options we have not investigated, but there are also a virtually limitless number of variations on the themes that we have investigated which remain unexplored. Modification of such options could generate proposals with a different scale or shape.

The second main section was much more qualitative: we did not attempt to cost any of these final regional options. Instead, we strove to highlight the way in which the uptake and use of renewable energy and energy-efficiency technologies might be increased by means of regional action. This kind of mitigation option has not, hitherto in the debate, received much attention, at least not within the context of the broader climate change discussions. Granted, issues such as free trade and collaborative research have certainly been investigated for other reasons. However, were they to be linked more directly to the climate change issue, then more arguments supporting them (which might, in turn, allow more resources to be leveraged) could be brought to bear. Regardless, we strived to show that the boundaries of potential regional options extend well beyond the electricity and transportation sectors, narrowly defined. Clearly, however, these investigations into other areas are at a much earlier stage. Nevertheless, they are sufficiently intriguing as to warrant pursuing.

ENDNOTES

1   The assistance of colleagues at the Centre for Energy, Environment, Science and Technology (CEEST) in Tanzania in formulating initial ideas about these other regional mitigation options is gratefully acknowledged.
2   This section builds upon work undertaken by Peter Zhou in the context of the African Energy Policy Research Network's (AFREPREN's) programme on energy and climate (see Zhou, 1997).
3   Indeed, recall (from Chapter 3) the value that SADCC placed upon effective regional transportation links.

4   In this chapter we examine the potential for regional cooperation among the 11 mainland countries of southern Africa (excluding, that is, the Democratic Republic of the Congo). We do this simply because it is for these countries that the best data are available.

5   Ideally, data for future energy demand would have originated from entities within the region itself. However, there are no comprehensive long-term national or regional transport plans. Although transport plans do exist at the national level, they usually have the same time horizon as those for national plans as a whole: this generally means ten years at most (though some countries have initiated National Development Visions which extend to approximately 20 years).

6   Though somewhat more conservative than the projections we used in estimating future electricity demand in Chapter 4, these figures are nevertheless still broadly consistent with the ones developed there.

7   Clearly, there are some problems with the estimates: Swaziland's transportation-related energy consumption, for example, is expected to decline between 1990 and 2010, in spite of positive population and GDP growth during the same period. Nevertheless, we believe that these are satisfactory estimates for our purposes.

8   Were other greenhouse gases to be considered as well, corresponding figures for carbon dioxide equivalence would be approximately 25 per cent higher.

9   For a general discussion of all kinds of options, see Michaelis (1996).

10   Remember that South Africa's main trading partners are located outside of the region: major export markets are located in Italy, Japan, the United States, Germany, the United Kingdom, other European Union countries and Hong Kong; while major imports are sourced from Germany, the United States, Japan, the United Kingdom and Italy (based on 1995 figures, taken from CIA, 1997).

11   One report noted that the total volume of cargo (excluding petroleum products) passing through Beira port in 1995 was 1500 kilotonnes, of which two-thirds originated in, or was destined for, Zimbabwe (*Cargo Info Africa: Freight & Trading Weekly*, 1997).

12   In this way, we are being consistent with the relationship portrayed in Figure 7.1.

13   We are taking an engine to weigh 96 tonnes, an empty wagon 20 tonnes, and we assume that each train is made up of 30 wagons.

14   The rehabilitation of the region's corridors could lessen the travel time on both rail and road. For the sake of exploring the issue, however, we assume the length of the return journey to be constant throughout the study period.

15   Strictly speaking, only 60 to 70 per cent of the cargo travelling along the corridor is regional (that is, it moves in both Mozambique and Zimbabwe). Nevertheless, we still examine all of it, accepting that we are overstating the mitigation potential (not all of the cargo will travel the entire length of the route).

16   Adapted from UNEP (1997). See, also, UNEP (forthcoming).

17   The key difference in this regard would be the reduction in the wear and tear of the regional road network, as well as the increased stress on the railway lines.

18   By ignoring this, we are assuming that the ongoing cost of replacing the two modes of transport – that is, trains and trucks – is approximately equal.

19   Full details of all options in this chapter are provided in UNEP (forthcoming).

20   For a broader analysis, see Ngwenya et al (1993).

21   Within southern African governments, the transportation and communications portfolios tend to be grouped together, while energy is usually bundled with mines, minerals or water in another ministry. Consequently, each group's action could be working to cross-purposes: transportation policies might be focused on providing the service without regard for the energy expended, while the energy

ministry might place restrictions on energy usage in the sector without due consideration for the provision of efficient transportation services.

22 'Tazara' comes from the name of one of the companies operating on this route: the Tanzania–Zambia Railway Authority.

23 Many of these other corridors may require substantial repairs to both road and rail before any further mitigation proposals could be seriously considered.

24 Though we are assuming that the cost of an electric train is the same as that of a diesel train, we are also assuming that the former can carry more cargo – 42 wagons instead of 30 (for the latter).

25 However, there is a great difference between a 25-MW hydropower facility (as envisaged here) and a 3000 MW one (as envisaged in Chapters 5 and 6)!

26 A full life-cycle analysis suggests that the emission reductions will be of the order of 10 to 30 per cent (Michaelis, 1996, p 695). For our calculations here, we will use a figure of 20 per cent.

27 As figures in this chapter have already revealed, South Africa is the main transportation demand centre in the region. By focusing solely upon South Africa, this fuel substitution option is not as explicitly regional as many of the others we explore. The sole regional characteristic is that the natural gas originates from outside of South Africa. Of course, this option could be extended to other countries of the region and a similar analysis undertaken.

28 This estimate follows from the fact that, in 1988, 62 per cent of South African carbon dioxide emissions 'from mobile consumption sources' came from cars and light trucks (Scholes and van der Merwe, 1995, p 10).

29 If we focused on greenhouse gas emissions, more broadly, we would undoubtedly pay much stricter attention to the potential for CNG leakage (particularly when filling vehicles). Given its significant global warming potential, even small leakages of CNG could negate any savings from reduced carbon dioxide emissions.

30 We are only including the cost of retrofitting the first generation of vehicles. By assuming that the 1 per cent penetration rate continues throughout the study period, we are also assuming that some kind of CNG-vehicle industry arises in order to replace (eventually) this first generation of retrofitted vehicles.

31 Additionally, the cost of establishing the distribution network for CNG should not be forgotten. Here, we assume that it will be city-based and therefore relatively modest. Given, however, the large land area of South Africa, a truly national network would be extremely expensive to construct.

32 Adapted from UNEP (1997). See also UNEP (forthcoming).

33 Though Angola has the potential to supply South Africa with its crude oil (and refined products), most actually comes from the Middle East.

34 For more on the state of the region's roads, see Mwase (1995a).

35 Indeed, the fact that the rehabilitation and maintenance of interstate highways has been a priority for, among others, the PTA/COMESA, further indicates that this mitigation option fits with existing regional priorities (for example, Mwase, 1995a, p 79).

36 These assumptions are as follows: cost of new truck is US$150,000; cost of new train is US$3,500,000; both trucks and trains operate 90 per cent of the time; and the cost of new pipeline is US$1000 per metre (thus, US$400 million for the length of the route under investigation).

37 Swaziland and Lesotho – the two other land-locked countries – are not only small demand centres but they are located relatively close to existing seaports and pipeline terminals. However, several countries that are serviced by pipeline might

still have some of their petroleum products brought in by other means: Zimbabwe, with 20 per cent of supply arriving by road or rail from South Africa, is a case in point ('Zimbabwe – Oil & Gas Industry Overview', 1998).

38 Regional markets may encourage new technologies: technologies need 'to reach sufficient volume to lower costs to become competitive' (Ishitani and Johansson, 1996, p 590).

39 Prior to this, the SADC(C) TAU had long been involved in analysing the potential for regional markets – in, for example, new and renewable energies (Raskin and Lazarus, 1991, p 170). See, also, the policy prescriptions in Karekezi and Ranja (1997, p 217).

40 Note that the mitigation gains in the Annex I group study of standards for refrigerators were projected to be most pronounced in the case of the Czech Republic (Mullins, 1996, p 10). Given that developing countries' societies more closely resemble such economies in transition than they do the industrialized countries, this encourages further investigation into this potential mitigation option.

41 It is worth noting, moreover, that training could be a means where regional action could advance mitigation goals and development ambitions with regard to biomass use (an issue of more salience to the local and national levels) and the region's woodfuel crisis more broadly. Cross-fertilization of ideas among different experts in the region (for example, those interested in agriculture, energy and forestry) might be facilitated by a regional entity – or at least regional activity – of some sort (following Raskin and Lazarus, 1991, p 168). This, of course, has been undertaken by, among others, SADC(C) in the past, but it could also be presented as a climate change mitigation option in the future.

Chapter 8 | Conclusions

Gordon A Mackenzie and Ian H Rowlands

This study has confirmed that regional options to mitigate climate change do indeed exist in southern Africa. Moreover, not only do these regional options exist, but some offer the potential for very considerable carbon dioxide reductions. Table 8.1 presents a summary of the options identified and quantified in terms of potential carbon dioxide abatement and cost. Figures in the second and third columns in the table indicate this sizeable potential for carbon dioxide reduction, while the figures in the final column suggest that these options might also be realizable at a competitive price.

For the sake of initial comparisons within the region, Table 8.2 lists the results of an assessment of national mitigation options from one of the most studied countries in the region, Zimbabwe. The abatement costs and potentials found in the Zimbabwe study are typical for countries in the region. Figure 8.1 shows a calculated abatement cost curve for Zimbabwe embodying some of these abatement options. Comparing the potential and cost of the regional options (Table 8.1) with those of the representative national options (see Figure 8.1) effectively illustrates the fact that regional options are potentially much larger than national options and cheaper than many of them as well. The existence of non-trivial, and conceivably afford-able, regional mitigation options is certainly encouraging.

This study has also confirmed, however, that the hurdles to successfully implementing regional options could prove considerable. Even if not insur-mountable, implementation barriers could certainly lessen the relative attractiveness of these regional options. In other words, all else being equal, regional mitigation options will probably be more difficult to put into practice than measures that involve only one country. Given the choice between a regional option (involving, say, five countries) that reduced carbon dioxide emissions by one million tonnes at a certain net cost, and a national option that achieved the same reductions at the same cost, it is almost certain that the national option would be preferred. The various factors that lend support to this assertion – those factors that prevent or hinder regional cooperative arrangements – were discussed in Chapter 2 and with particular reference to the southern African region in Chapter 3. Indeed, our explo-ration of implementation issues – particularly in Chapter 6 – gathered empirical material that tends to confirm this. The importance of sovereignty, security and national interests more broadly, the need to ensure an equitable

Table 8.1 *Regional options to mitigate global climate change*

Description of option	Total carbon dioxide abated over the study period (effectively 2005 to 2050) (million tonnes)	Average annual carbon dioxide abated § (million tonnes)	Estimated average abatement cost (US dollars per tonne carbon dioxide)*
Extensive use of hydropower: hydrophilic scenario	4,755	108	41.80
Moderate use of hydropower and gas: hybrid scenario	1,971	45	6.26
500 MW hydropower facility on the Zambezi River	137	3.0	–2.04
3000 MW hydropower facility in the DRC	823	17.9	–2.46
500 MW gas-fired power station in Mozambique	56	1.2	16.71
Moving from road to rail along the Beira–Mutare route	0.67	0.015	8.58
Electrifying the railway along the Beira–Mutare route	1.5	0.032	38.28
Electrifying the railway along the Beira–Mutare route and moving from road to rail	2.55	0.055	27.57
Fuel switching from gasoline to CNG: 1 per cent of cars in South Africa	5.4	0.12	1.37
Building a pipeline from Pretoria to Gaborone	2.6	0.057	63.41

§ The average annual amount of carbon dioxide abated is calculated as the total amount abated divided by either 44 years (abatement begins to take effect in the year 2007 for the two scenarios) or 46 years (abatement begins to take effect in the year 2005 for the remaining projects).
* The average abatement cost is calculated as the net present value of the abatement scenario or option, less the net present value of the baseline scenario or option, at a discount rate of 5 per cent.

Table 8.2 *Carbon dioxide abatement options in Zimbabwe,*
cost and potential in 2030

Reduction option	Annualized abatement cost (US dollars per tonne carbon dioxide)	Carbon dioxide reduction in 2030 (million tonnes)
Ethanol blend	−186.49	0.044
Tillage	−134.96	0.022
Cokeoven gas for Hwange	−54.87	0.043
Efficient lighting	−31.16	0.000
Geyser timeswitches	−14.05	0.116
Coalbed ammonia	−11.97	0.800
Methane from sewage	−11.88	0.024
Prepayment meters	−9.34	0.006
Efficient refrigerators	−8.90	0.110
Efficient motors	−8.38	0.264
Efficient tobacco barns	−6.83	0.418
Efficient boilers	−4.68	2.083
Pine afforestation	0.12	2.942
Hydropower	1.86	3.662
Biogas from landfills	2.75	0.447
Biogas for rural households	4.74	0.094
Efficient furnaces	9.11	0.825
Solar geysers	29.05	0.258
Central PV electricity	44.19	0.411
Power factor correction	555.00	0.658
Solar PV water pumps	1,863.74	0.000

Note: 5 per cent discount rate employed
Source: adapted from UNEP, 1993

distribution of costs and benefits (or at least a perception of the same), and the need to manage the often fractious relationship between South Africa and the rest of the region are but some examples of the difficulties that could well be part and parcel of most regional mitigation options. Though we proposed some ideas on how these barriers might themselves be mitigated, we have only begun to scratch the surface of this issue.

It is also worth recognizing, however, that there could conceivably be forces and factors to encourage – rather than to discourage – successfully implementing regional strategies, mitigation options included. The 'world community', for one, has voiced its support for regional arrangements in different parts of the developing world, and it is hoped that international resources will be marshalled to support any regional strategy. This is not meant to suggest that extraregional actors will be a panacea of any kind. Instead, it serves as a reminder that this extraregional interest could be mobilized in support of broader regional aims.

What, however, will probably be decisive to the prospects for implementation are the non-climate consequences of the mitigation options – what

Abatement cost (US dollars/tonne)

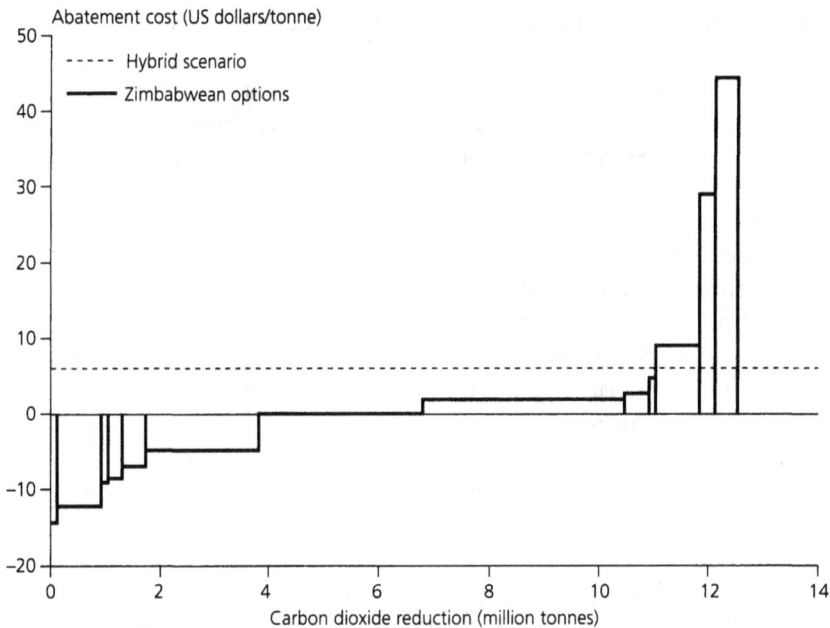

Notes: The vertical axis indicates the annualized cost of abatement and the horizontal axis represents the amount of carbon dioxide reduced in the year 2030. The dotted line shows the approximate annual average abatement cost of the hybrid scenario, indicating that this set of regional mitigation actions may be cheaper than moderately priced options at the national level.
Source: Tables 8.1 and 8.2

Figure 8.1 *Abatement cost curve for Zimbabwe for 2030, showing selected national options with one regional option superimposed*

we have called the developmental consequences. If clearly positive, entities in the region will be motivated to support implementation; if neutral, or even only marginally positive, 'champions' for the mitigation options – if we may use that term – will not arise, and the climate goals will inevitably prove difficult to realize. Though a well-worn and often-misused term, win–win nevertheless seems to capture the sentiment.

At least some of our options were characterized in this way: they appeared to offer not only significant climate change benefits at reasonable prices, but they also had the potential to generate a range of non-climate benefits for the region. Given this, we are convinced that regional mitigation options warrant further attention. This should not be interpreted as unconditional praise for regional mitigation options, nor is it meant to imply that regional options are preferable. Instead, it suggests continued curiosity, coupled with cautious optimism, that regional mitigation options could be part of a fair and efficient portfolio of climate change responses. Indeed, an important conclusion of the study is that some regional options may be attractive, and that they should definitely be considered along with the national measures. To determine how this might be done, however, a number of issues clearly still need to be addressed.

For one, we have only begun to explore the appropriate methodology for a regional study. Though we have attached numbers (abatement potential and incremental cost) to a selection of representative mitigation options by comparing them with a regional baseline, they only represent a first approximation. Nevertheless, the guidance offered by the methodology for national studies (see Chapter 1) appeared to be appropriate. We developed a baseline, assuming this to be a future development scenario of the region's countries in a situation where no explicit priority is given to climate change mitigation, whether at the national or the regional level. This choice of baseline scenario allows national options and regional (or multinational) options to be assessed and compared with each other.

More work, however, needs to be completed with respect to integrating national and regional options – something that we have not attempted to do (apart from the preliminary comments earlier in this chapter). We recognize that, in moving from a purely national to a multicountry or regional perspective, the potential and cost of national options will change. In other words, the acceptance, or opening up, of bilateral or multilateral cooperation with respect to mitigation options would have the effect of changing the potential amount of greenhouse gas abated and the cost of measures that could otherwise be implemented by the countries in isolation.

Nevertheless, we maintain that the technical aspects of climate change mitigation projects at the regional level need not be qualitatively different from nationally implemented measures. Moreover, we suggest that the methods for analysing them – the concepts of baseline, mitigation scenario and the like – require, in principle, little modification from the single country case. It is when we consider implementing regional mitigation options that more significant issues present themselves. Consequently, development of a comprehensive methodology for regional mitigation analysis still needs to be completed.

This study was by no means exhaustive with regard to the coverage of mitigation options in the region. A complete survey and analysis of the potential to reduce greenhouse gas emissions (or to enhance sinks) was beyond the scope of the work. Instead, we focused on a few options, with most detail in the power sector. For many reasons, electricity supply is quite appropriate for such an initial study, not least because of the significant greenhouse gas emissions characterizing it, as well as the regional cooperation that is already evident. Nevertheless, other options merit much more careful consideration.

The group of countries in the Southern African Development Community (SADC) was chosen as the subject of study, and the options are naturally specific to that region. However, we believe that the essential message to be gained from this study has relevance beyond the region, and indeed can be applied worldwide. Therefore, another conclusion is that other regions of the world might fruitfully consider regional approaches: it appears to be possible within the terms of the UN Framework Convention on Climate Change (UNFCCC) – in terms of both a response to climate change (that is, mitigation or adaptation activity) and a political action (that is, participation in the UNFCCC and broader elements of the regime).

This study provides support for participation of regional entities in climate fora. More specifically, we hope that it will feed into policy discussions and sensitize policy-makers to the possibilities of mitigation options at the regional level. Returning to the region in focus, a central message of the study is that southern African decision-makers should consider explicitly the significant global environmental resources that are available for carbon dioxide abatement. Considering regional mitigation options only widens the menu of options that might attract international support.

In summary, if the barriers to implementation can be overcome, then the potential of regional mitigation options could be exploited to the benefit of the region, the member countries and the global community as a whole. These options could be consistent with national developmental priorities and, indeed, enhance the developmental prospects of the nations involved. In other words, reaching beyond national boundaries to explore sustainable futures that require regional cooperation may lead to situations that will improve sustainability at all levels.

References

Activities Implemented Jointly (1996) 'Activities Implemented Jointly: Annual Review of Progress Under the Pilot Phase', FCCC/CP/1996/14, 4 June

Advanced Resources International, Inc and Resource Exploration and Development (1995) 'Coalbed Methane: A Promising New Gas Resource for Southern Africa', paper presented at Sub-Saharan Oil and Minerals Conference, 27–29 March

AfDB (1996) *The Electricity Sub-Sector in Africa*, Abidjan: African Development Bank, Energy Technical Paper No ES1

AfDB (1993a) *Economic Integration in Southern Africa, Executive Summary*, Abidjan: African Development Bank

AfDB (1993b) *Economic Integration in Southern Africa*, volume 2, Abidjan: African Development Bank

Africa South of the Sahara (1995) twenty-fourth edition, London: Europa Publications Ltd

Allison, Graham T (1971) *Essence of Decision: Explaining the Cuban Missile Crisis*, Boston, MA: Little, Brown and Company

Aly, Ahmad A H M (1994) *Economic Cooperation in Africa: In Search of Direction*, London: Lynne Rienner Publishers

'America Loses its Afrophobia' (1997) *The Economist*, 26 April, pp 47–48

Anglin, Douglas G (1983) 'Economic Liberation and Regional Cooperation in Southern Africa: SADCC and PTA', *International Organization*, vol 37, no 4, Autumn pp 681–711

Annex I Expert Group (1996) *Sustainable Transport Policies: CO_2 Emissions from Road Vehicles*, Paris: IEA/OECD

Arrow, K J, Parikh, J and Pillet, G (1996) 'Decision-Making Frameworks for Addressing Climate Change' in IPCC, *Climate Change 1995: Economic and Social Dimensions of Climate Change*, Cambridge: Cambridge University Press, pp 55–77

Asante, S K B (1995) 'The Need for Regional Integration: A Challenge for Africa', *Review of African Political Economy*, vol 22, no 66, December, pp 573–77

Axline, W Andrew (1994a) 'Comparative Case Studies of Regional Cooperation Among Developing Countries' in W Andrew Axline (ed) *The Political Economy of Regional Cooperation: Comparative Case Studies*, London: Pinter Publishers, pp 7–33

Axline, W Andrew (1994b) 'Cross-Regional Comparisons and the Theory of Regional Cooperation: Lessons from Latin America, the Caribbean, South East Asia and the South Pacific' in W Andrew Axline (ed) *The Political Economy of Regional Cooperation: Comparative Case Studies*, London: Pinter Publishers, pp 178–224

Axline, W Andrew (1979) *Caribbean Integration: The Politics of Regionalism*, London: Frances Pinter Ltd

Bacon, Robert W, Besant-Jones, John E and Heidarian, Jamshid (1996) *Estimating Construction Costs and Schedules: Experience with Power Generation Projects in Developing*

Countries, Washington, DC: World Bank Technical Paper no 325, Energy Series, August

Balassa, Bela (1961) *The Theory of Economic Integration*, Homewood, IL: Richard D Irwin, Inc

Banuri, T, Göran-Mäler, K, Grubb, M, Jacobson, H K and Yamin, R (1996) 'Equity and Social Considerations' in IPCC, *Climate Change 1995: Economic and Social Dimensions of Climate Change*, Cambridge: Cambridge University Press, pp 83–124

Batoka Joint Venture Consultants (1993) *Batoka Gorge Hydro Electric Scheme Feasibility Study*, Lahmeyer International Consulting Engineers and Knight Piesold Consulting Engineers, Final Report, September

Beenstock, Michael, Goldin, Ephraim and Haitovsky, Yoel (1997) 'The Cost of Power Outages in the Business and Public Sectors in Israel: Revealed Preference vs Subjective Valuation', *The Energy Journal*, vol 18, no 2, pp 39–61

Besant-Jones, John (1996) 'Guidelines for Attracting Developers of Hydropower Independent Power Projects', *Energy Issues: Energy Note*, Washington, DC: The World Bank Group, no 9, April

Besant-Jones, John (1995) 'Attracting Finance for Hydroelectric Power', *Energy Issues: FPD Energy Note*, Washington, DC: The World Bank Group, no 3, June

Black and Veatch International (1996a) *Hwange Thermal Power Plant Expansion, Project Document*, vol 1, Kansas City, MI: Black and Veatch International, January

Black and Veatch International (1996b) *Mchuchuma/Katewaka Coal-Fired Power Plant Feasibility Study*, Arlington, VA: Black and Veatch International, 22 November

Blomqvist, Hans (1993) 'ASEAN as a Model for Third World Regional Economic Co-operation?', *ASEAN Economic Bulletin*, vol 10, no 1, July, pp 52–67

Blumenfeld, Jesmond (1991) *Economic Interdependence in Southern Africa: From Conflict to Cooperation?*, London: Pinter Publishers for the Royal Institute of International Affairs

Bolin, Bert and Houghton, John (1995) 'Berlin and Global Warming Policy', *Nature*, vol 375, no 6528, 18 May, p 176

BPC (1997) personal communication, 'BPC Interconnected Power System Operating Statistics', 14 January

British Petroleum (1997) *The BP Statistical Review of World Energy*, London: British Petroleum

Brown, M Leann (1994) *Developing Countries and Regional Economic Cooperation*, London: Praeger

BTP (1993) *Botswana Transport Plan, (Stage II)*, Gaborone: Department of Roads, Ministry of Works, Transport and Communications, Government of Botswana

Calland, Richard and Weld, David (1994) *Multilateralism, Southern Africa and the Postmodern World: An Exploratory Essay*, Southern African Perspectives, Working Paper no 38, Centre for Southern African Studies, University of the Western Cape

Cantwell, John (1991) 'Foreign Multinationals and Industrial Development in Africa' in Peter J Buckley and Jeremy Clegg (eds) *Multinational Enterprises in Less Developed Countries*, Basingstoke: Macmillan, pp 183–224

Cargo Info Africa: Freight & Trading Weekly (1997) 'Port of Beira Handles More Cargo', http://www.rapidttp.com/cargo/ftw/97/97fe14g.html; 14 February; accessed 1 March 1998

Carrim, Yasmin (1994) *The Preferential Trade Area for Eastern and Southern Africa and COMESA: A Call for Suspension*, Bellville: Centre for Southern African Studies, Southern African Perspectives: A Working Paper Series, no 32

Charpentier, J P and Schenk, K (1995) 'International Power Interconnections: Moving from Electricity Exchange to Competitive Trade', *Public Policy for the Private Sector*,

Washington, DC: World Bank, Industry and Energy Department, Note no 42, March

CIA (1997) *The World Factbook, 1997*, Washington, DC: Central Intelligence Agency

CIA (1996) *The World Factbook 1996*, Central Intelligence Agency, http://www.odci.gov/cia/publications/nsolo/wfb-all.htm; accessed 10 January 1998

'Coal-Fired South Africa Makes Room for Renewable Energy' (1996) *World Rivers Review*, vol 11, no 5, December

COMESA Treaty (1994) 'Common Market for Eastern and Southern Africa: Treaty Establishing', reprinted in *International Legal Materials*, vol 33, pp 1067ff

'Comment' (1997) *Southern African Economist*, July-August, p 2

Cortell, Andrew P and Davis, Jr, James W (1996) 'How Do International Institutions Matter? The Domestic Impact of International Rules and Norms', *International Studies Quarterly*, vol 40, no 4, December, pp 451–78

Curry Jr, Robert L (1991) 'Regional Economic Co-operation in Southern Africa and Southeast Asia', *ASEAN Economic Bulletin*, vol 8, no 1, July, pp 15–27

Dale, A P (1995) 'An Energy Sector Overview of Zambezi Basin Developments' in T Matiza, S Crafter and P Dale (eds) *Water Resource Use in the Zambezi Basin: Proceedings of a Workshop Held at Kasane, Botswana, 28 April – 2 May 1993*, Gland: IUCN, pp 147–52

Dash, Kishore C (1995) 'Swords into Ploughshares: Challenges for Regional Cooperation in South Asia', *Asian Profile*, vol 23, no 6, December, pp 511–26

Datta, Ansu (1989) 'Strategies for Regional Cooperation in Post-Apartheid Southern Africa – the Role of Non-Governmental Organizations' in Bertil Odén and Haroub Othman (eds) *Regional Cooperation in Southern Africa: A Post-Apartheid Perspective*, Uppsala: The Scandinavian Institute of African Studies, pp 91–102

de Melo, Jaime and Panagariya, Arvind (1993) 'Introduction' in Jaime de Melo and Arvind Panagariya (eds) *New Dimensions in Regional Integration*, Cambridge: Cambridge University Press, pp 3–21

Deutsch, K W (1957) *Political Community and the North Atlantic Area*, Princeton, NJ: Princeton University Press

Directorate of Roads (1995) *Transport Statistics 1995*, Pretoria: South Africa Department of Transport, Report NSC 18/95

Dludlu, John (1997) 'Comesa Attempts to Recapture Two Eloping Member Nations', *Business Day*, 15 January

Dorcey, Tony (1997) *Large Dams: Learning from the Past, Looking at the Future*, proceedings from a workshop held in Gland, Switzerland, 11–12 April 1997, Gland: IUCN

Dutkiewicz, R K (1996) 'Energy Demand and Supply in Sub-Equatorial Africa', *Journal of Energy in Southern Africa*, vol 7, no 3, August 1996, pp 73–83

EdF (1997) 'Grand Inga', personal communication from Richard Rolland du Roscoat, Paris: Electricité de France, 14 August

EIU (1997) *Namibia: Country Report*, London: Economist Intelligence Unit, Third Quarter

EIU (1996a) *Zambia Country Profile 1996–97*, London: Economist Intelligence Unit

EIU (1996b) *Zambia: Country Report*, London: Economist Intelligence Unit, Second Quarter

EIU (1996c) *Zimbabwe: Country Report*, London: Economist Intelligence Unit, Fourth Quarter

'Energy White Paper Spells Out Focus' (1998) *The Namibian Economist*

Engineering News (1997) 23–29 May

Escom (1997) personal communication from L J Nchembe, Projects Manager, 'Greenhouse Gas Mitigation Under SAPP', 1 September

Eskom (1998) 'Products & Services – Transmission – Join ... res – Major Projects in
    Southern Africa', http://www.eskom.co.za/text/products/trans/sapp/majprj.htm;
    accessed 15 February
Eskom (1996a) *Statistical Yearbook 1995*, Johannesburg: Eskom
Eskom (1996b) *Synopsis of the Integrated Electricity Plan for Eskom's 1997–2001 Business
    Cycle*, Johannesburg: Eskom
Eskom (1995) *Environmental Report*, Johannesburg: Eskom
ETSU (1995) *Mechanisms for Improved Energy Efficiency in Transport*, London: ODA
    Report, ETSU/RYCA/18400304/Z3/Issue 2
EU (1994) 'Declaration by the EU/Southern Africa – Ministerial Conference of 5/6
    September 1994 in Berlin', 6 September 1994, released by the European Union
    on 28 September
'Focus on Energy-Efficiency Options' (1997) *Business Day*, 5 August
Foroutan, Faezeh (1993) 'Regional Integration in Sub-Saharan Africa: Past Experience
    and Future Prospects' in Jaime de Melo and Arvind Panagariya (eds) *New
    Dimensions in Regional Integration*, Cambridge: Cambridge University Press,
    pp 234–271
Forsius, John (1993) 'Rapporteur's Report on Session 1, Regulatory Challenges
    Facing Hydropower Today' in *Hydropower, Energy and the Environment: Options for
    Increasing Output and Enhancing Benefits, Conference Proceedings*, conference in Sweden,
    June 1993; Paris: International Energy Agency
Freer, Gordon (1996) *Southern Africa, Sustainable Development and South-South Cooperation*,
    Centre for Southern African Studies, no 51
Gagnon, Luc and van de Vate, Joop F (1997) 'Greenhouse Gas Emissions from
    Hydropower: The State of Research in 1996', *Energy Policy*, vol 25, no 1, 1997,
    pp 7–13
Gambari, Ibrahim A (1991) *Political and Comparative Dimensions of Regional Integration: The
    Case of ECOWAS*, London: Humanities Press International, Inc
Garribba, Sergio (1993) 'Opening Address for the International Conference on
    Hydropower, Energy and the Environment' in *Hydropower, Energy and the Environment:
    Options for Increasing Output and Enhancing Benefits, Conference Proceedings*, conference in
    Sweden, June 1993; Paris: International Energy Agency, pp 27–35
George C Marshall Institute (1998) *Are Human Activities Causing Global Warming?*,
    http://www.marshall.org/Warming.html; accessed 1 March
Gibb, Richard (1997) 'Regional Integration in Post-Apartheid Southern Africa: The
    Case of Renegotiating the Southern African Customs Union', *Journal of Southern
    African Studies*, vol 23, no 1, March, pp 67–86
Gibb, Richard (1996) 'Towards a New Southern Africa: Regional Challenges', *The
    South African Journal of International Affairs*, vol 4, no 1, Summer, pp 1–26
Gibb, Richard (1994) 'Regional Economic Integration in Post-Apartheid Southern
    Africa' in Richard Gibb and Wieslaw Michalak (eds) *Continental Trading Blocs: The
    Growth of Regionalism in the World Economy*, Chichester: John Wiley and Sons,
    pp 208–229
Goldsmith, E and Hildyard, N (1984) *The Social and Environmental Effects of Large Dams*,
    vol I: Overview, Camelford: Wadebridge Ecological Centre
Goodland, Robert (1995) *'The Big Dams Controversy': Killing Hydro Promotes Coal and
    Nukes: Is that Better for Environmental Sustainability?*, Washington, DC: World Bank,
    Public Lecture in GTE Technology and Ethics Series, 8 May
'The Great Inga Dream' (1995) *African Business*, no 198, April, p 28
Green, Reginald Herbold (1990) 'Economic Integration/Coordination in Africa: The
    Dream Lives But How Can it be Lived?' in James Pickett and Hans Singer (eds)

Towards Economic Recovery in Sub-Saharan Africa, London: Routledge, pp 106–28

Gumende, António (1996a) 'Can SADC Deliver?', *SAPEM*, August, pp 5–8

Gumende, António (1996b) 'The Bully on the Bloc', *SAPEM*, August, pp 8–9

Gumende, António (1996c) 'The Quest for Markets Access', *SAPEM*, August, pp 23–24

Haarlov, Jens (1997) *Regional Cooperation and Integration within Industry and Trade in Southern Africa: General Approaches, SADCC and the World Bank*, Aldershot: Avebury

Haas, Ernst B (1990) *When Knowledge is Power: Three Models of Change in International Organizations*, Berkeley, CA: University of California Press

Haas, Ernst B (1958) *The Uniting of Europe: Political, Social and Economic Forces, 1950–1957*, London: Stevens

Haas, Peter M (1990) *Saving the Mediterranean: The Politics of International Environmental Cooperation*, New York: Columbia University Press

Haites, Erik F and Rose, Adam (1996) 'Energy and Greenhouse Gas Mitigation: The IPCC Report and Beyond', special issue of *Energy Policy*, vol 24, Nos 10/11, October/November

Hansen, Roger D (1969) 'Regional Integration: Reflections on a Decade of Theoretical Efforts', *World Politics*, vol 21, no 2, January, pp 242–71

Hardin, Garrett (1968) 'The Tragedy of the Commons', *Science*, vol 162, 13 December, pp 1243–48

Harvey, L D Danny and Bush, Elizabeth J (1997) 'Joint Implementation: An Effective Strategy for Combating Global Warming?', *Environment*, vol 39, no 8, October, pp 15–20, 36–44

Hawkins, Anthony M (1992) 'Economic Development in the SADCC Countries' in Gavin Maasdorp and Alan Whiteside (eds) *Towards a Post-Apartheid Future: Political and Economic Relations in Southern Africa*, Basingstoke: Macmillan, pp 105–31

Hawkins, Tony (1994) *The New South Africa: Business Prospects and Corporate Strategies*, London: Economist Intelligence Unit, August

Heggie, Ian G (1995) *Management and Financing of Roads: An Agenda for Reform*, Washington, DC: World Bank Technical Paper Number 275, Africa Technical Series

Hettne, Bjorn and Inotai, Andras (1994) *The New Regionalism: Implications for Global Development and International Security*, Helsinki: UNU WIDER

Hulme, Mike (ed) (1996) *Climate Change and Southern Africa: An Exploration of Some Potential Impacts and Implications in the SADC Region*, Norwich: Climate Research Unit and WWF

ICOLD (1997) *1996 Annual Report*, Paris: International Commission on Large Dams

IFC (1995) *Emerging Stock Markets Factbook 1995*, Washington, DC: International Finance Company

IMF (1997) *Direction of Trade Statistics Quarterly*, Washington, DC: International Monetary Fund, June

IPCC (1996a) *Climate Change 1995: The Science of Climate Change*, Cambridge: Cambridge University Press

IPCC (1996b) *Climate Change 1995: Impacts, Adaptations and Mitigation of Climate Change: Scientific-Technical Analyses*, Cambridge: Cambridge University Press

IPCC (1990) *Energy and Industry Subgroup Report*, Energy and Industry Subgroup, 31 May

Ishitani, H and Johansson, T B (1996) 'Energy Supply Mitigation Options' in IPCC, *Climate Change 1995: Impacts, Adaptations and Mitigation of Climate Change: Scientific-Technical Analyses*, Cambridge: Cambridge University Press, pp 587–647

Jackson, Robert H (1990) *Quasi-States: Sovereignty, International Relations, and the Third World*, Cambridge: Cambridge University Press

Jefferis, K R (1995) 'The Development of Stock Markets in Sub-Saharan Africa', *South African Journal of Economics*, vol 63, pp 346–63

Jinadu, L Adele (1990) 'Regional Integration in Africa: Theoretical Perspectives and Their Implications', *Eastern Africa Social Science Research Review*, vol 6, no 1, January, pp 1–12

Johnson, Omotunde E G (1991) 'Economic Integration in Africa: Enhancing Prospects for Success', *Journal of Modern African Studies*, vol 29, no 1, March, pp 1–28

Jordanger, Einar (1992) 'Power Cooperation in the Southern African Region', *SADCC Energy*, vol 9, no 23, pp 11–17

Kapata, Dennis (1997) 'Can Zimbabwe Hit Back?', *Southern African Economist*, February, p 23

Karekezi, Stephen and Ranja, Timothy (1997) *Renewable Energy Technologies in Africa*, London: Zed Books

Katerere, Yemi, Moyo, Sam and Ngobese, Peter (nd) 'Opportunities for NGO Involvement in the Southern African Development Community's (SADC) Strategies on the Environment', Harare: ZERO

Kayaya, Musengwa (1997) 'Comesa Secretary General, Mutharika, Suspended', *Panafrican News Agency*, 18 January

Keet, Dot (1996) 'The European Union's Proposed Free Trade Agreement with South Africa', *SAPEM*, September, pp 42–46

Keohane, Robert O (1984) *After Hegemony: Cooperation and Discord in the World Political Economy*, Princeton, NJ: Princeton University Press

Keohane, Robert O and Nye, Joseph S (1977) *Power and Interdependence: World Politics in Transition*, Boston, MA: Little, Brown & Company

Kibble, Steve, Goodison, Paul and Tsie, Balefi (1995) 'The Uneasy Triangle: South Africa, Southern Africa and Europe in the Post-Apartheid Era', *International Relations*, vol 12, no 4, April, pp 41–61

Kindleberger, Charles P (1986) 'International Public Goods Without International Government', *The American Economic Review*, vol 76, no 1, March, pp 1–13

Kisanga, Eliawony J (1991) *Industrial and Trade Cooperation in Eastern and Southern Africa*, Aldershot: Avebury

Kyoto Protocol (1997) 'Kyoto Protocol to the United Nations Framework Convention on Climate Change', FCCC/CP/1997/L.7/Add.1, 10 December

Langhammer, Rolf J (1991) 'ASEAN Economic Co-operation: A Stock-Taking From a Political Economy Point of View', *ASEAN Economic Bulletin*, vol 8, no 2, November

Langhammer, Rolf J and Hiemenz, Ulrich (1990) *Regional Integration Among Developing Countries: Opportunities, Obstacles and Options*, Tubingen: J C B Mohr

Laszlo, Ervin with Kurtzman, Joel and Bhattacharya, A K (1981) *RCDC Regional Cooperation Among Developing Countries): The New Imperative of Development in the 1980s*, Oxford: Pergamon Press

Lennon, S J (1996) 'The Role of Global Environmental Issues in Technology Choices Relating to the Future Supply and Demand of Electricity in Southern Africa' in S J Lennon et al, *Report on WEC Study 'Global Energy Perceptions to 2050 and Beyond for the African Region'*, Sandton: Eskom Technology Group, October

Lennon, S J et al (1996) *Report on WEC Study 'Global Energy Perceptions to 2050 and Beyond for the African Region'*, Sandton: Eskom Technology Group, October

Leys, Roger and Tostensen, Arne (1982) 'Regional Cooperation in Southern Africa: The Southern African Development Coordination Conference', *Review of African*

Political Economy, no 23, January–April, pp 52–71

Lourens, Carli (1997) 'Coal to Remain Chief SA Power Fuel Beyond 2000', *Martin Creamer's Engineering News*, 30 May

Lwiindi, R S (nd) 'An Overview of the Future Zambian Power System', *mimeo*, Lusaka: ZESCO

Maasdorp, Gavin (1992) *Economic Cooperation in Southern Africa: Prospects for Regional Integration*, London: Conflict Studies Paper no 253, July/August

Magadza, C H D (1994) 'Climate Change: Some Likely Multiple Impacts in Southern Africa', *Food Policy*, vol 19, no 2, pp 165–91

Majot, Juliette (1996) 'Risky Business', *World Rivers Review*, January

Mandaza, Ibbo, Mudenda, Gilbert and Chipeta, Chinyamata (1994) *The Joint PTA/SADCC Study: On Harmonisation, Rationalisation and Coordination of the Activities of the Preferential Trade Area for Eastern and Southern African States (PTA) and the Southern African Development Community (SADC)* July

Maya, R S (1982) 'Transitions in Zimbabwe's Hydroelectric Energy Base', *mimeo*

Maya, R S and Gupta, J (1996) *Joint Implementation: Carbon Colonies or Business Opportunities? Weighing the Odds in an Information Vacuum*, Harare: Southern Centre Publications

McCully, Patrick (1996) *Silenced Rivers: The Ecology and Politics of Large Dams*, London: Zed Books

Merz and McLennan (1981) *Report on Power Development Plan*, vol 2 of 4, main report, Newcastle upon Tyne, December

Meyer, Henrik et al (1996) *Assessment of Environmental External Effects in Power Generation*, Roskilde: Risø National Laboratory, Risø-R–938(EN), December

Meyns, Peter (1984) 'The Southern African Development Coordination Conference, (SADCC) and Regional Cooperation in Southern Africa' in Domenico Mazzeo (ed) *African Regional Organizations*, Cambridge: Cambridge University Press, pp 196–224

Michaelis, Laurie (1996) 'Mitigation Options in the Transportation Sector' in IPCC, *Climate Change 1995: Impacts, Adaptations and Mitigation of Climate Change: Scientific-Technical Analyses*, Cambridge: Cambridge University Press, pp 679–712

Milner, Helen (1992) 'International Theories of Cooperation Among Nations: Strengths and Weaknesses', *World Politics*, vol 44, no 3, April, pp 466–96

Ministry of Mines and Energy (1997) personal communication on 'Power Demand Projections in Namibia' from J V Mazeingo, Acting Permanent Secretary, Ministry of Mines and Energy, Republic of Namibia, 14 January

Ministry of Natural Resources and Energy (1997) personal communication on 'Electricity Demand Projections' from H D Shongwe for the Principal Secretary, Ministry of Natural Resources and Energy, Swaziland Government, Mbabane, 29 January

Moreira, José Roberto and Poole, Alan Douglas (1993) 'Hydropower and its Constraints' in Thomas B Johansson et al (eds) *Renewable Energy: Sources for Fuels and Electricity*, London: Earthscan, pp 73–119

Mukosa, C, Pitchen, G and Cadou, C (1995) 'Recent Hydrological Trends in the Upper Zambezi and Kafue Basins' in T Matiza, S Crafter and P Dale (eds) *Water Resource Use in the Zambezi Basin: Proceedings of a Workshop Held at Kasane, Botswana, 28 April – 2 May 1993*, Gland: IUCN, pp 85–98

Mullins, Fiona (1996) *Demand Side Efficiency: Energy Efficiency Standards for Trade Products*, Paris: Annex I Expert Group on the UNFCCC, 'Proposals and Measures for Common Action', Working Paper 5

Musoke, Issaa K S (1990) 'Beyond Summits and Jet-Setting: Prospects for Regional Integration in Southern Africa', *Eastern Africa Social Science Research Review*, vol 6, no 1, January, pp 30–47

Mwase, Ngila (1995a) 'Road and Bridge Maintenance Strategies in Eastern and Southern Africa: Opportunities and Challenges', *International Journal of Transport Economics*, vol 22, no 1, February, pp 65–84

Mwase, Ngila (1995b) 'Economic Integration for Development in Eastern and Southern Africa: Assessment and Prospects', *The Round Table*, no 336, October, pp 477–93

Mwase, Ngila (1994) 'Economic Integration for Development in Eastern and Southern Africa: Assessment and Prospects', *IDS Bulletin*, vol 25, no 3, pp 31–39

Mytelka, Lynn K (1994) *South–South Cooperation in a Global Perspective*, Paris: Organization for Economic Cooperation and Development

National Communications (1997) 'National Communications: Communications from Parties Included in Annex I to the Convention, First Compilation and Synthesis of Second National Communications from Annex I Parties, Addendum', FCCC/SBI/1997/19/Add.1, 9 October

Ngwenya, Sindiso et al (eds) (1993) *The Transport and Communications Sector in Southern Africa*, Harare: SAPES Books

North, Douglass C (1990) *Institutions, Institutional Change and Economic Performance*, Cambridge: Cambridge University Press

Nye, J S (1965) 'Patterns and Catalysts in Regional Integration', *International Organization*, vol 19, Autumn, pp 870–84

Odén, Bertil (1993) 'Introduction' in Bertil Odén (ed) *Southern Africa After Apartheid: Regional Integration and External Resources*, Uppsala: Nordiska Afrikainstitutet, pp 11–23

OECD (1997) *Geographical Distribution of Financial Flows to Aid Recipients: Disbursements, Commitments, Country Indicators, 1991–1995*, Paris: Organization for Economic Cooperation and Development

OECD (1993) *Regional Integration and Developing Countries*, Paris: Organization for Economic Cooperation and Development

Osherenko, Gail and Young, Oran R (1989) *The Age of the Arctic: Hot Conflicts and Cold Realities*, Cambridge: Cambridge University Press

Østergaard, Tom (1993) 'Classical Models of Regional Integration – What Relevance for Southern Africa?' in Bertil Odén (ed) *Southern Africa After Apartheid: Regional Integration and External Resources*, Uppsala: Nordiska Afrikainstitutet, pp 27–47

'Our Portfolio Poll' (1997) *The Economist*, 19 April, p 80

Park, Jong K (1995) 'The New Regionalism and Third World Development', *Journal of Developing Societies*, vol 11, no 1, pp 21–35

Paxton, Brian (1997) 'Southern African Fuel Supply and Demand Outlook', paper presented at UNCTAD Oil Conference, Harare

Raskin, P and Lazarus, M (1991) 'Regional Energy Development in Southern Africa: Great Potential, Great Constraints', *Annual Review of Energy and Environment*, vol 16, pp 145–78

Ravenhill, John (1979) 'Regional Integration and Development in Africa: Lessons from the East African Community', *Journal of Commonwealth and Comparative Politics*, vol 17, no 3

Redeby, L M Khalema et al (1994) *Planning and Management of the Electric Power Sector: Implications for Meeting Industrial Electricity Demand*, Maseru: Lesotho Electricity Corporation, for AFREPREN, Electricity Theme Group, June

Report of the Conference of the Parties (1995) 'Report of the Conference of the Parties on its First Session, Held at Berlin from 28 March to 7 April 1995, Addendum', FCCC/CP/1995/7/Add.1, 6 June

Robson, Peter (1990) 'Economic Integration in Africa: A New Phase?' in James Pickett and Hans Singer (eds) *Towards Economic Recovery in Sub-Saharan Africa*, London: Routledge, pp 129–49

Robson, Peter (1980) *The Economics of International Integration*, London: George Allen and Unwin

Rosenberg, D M, Berkes, F, Bodaly, R A, Hecky, R E, Kelly, C A and Mudd, J W M (1997) 'Large-scale Impacts of Hydroelectric Development', *Environmental Reviews*, vol 5, no 1, pp 27–54

Rowlands, Ian H (forthcoming) 'Encouraging Regional Sustainable Development: A Role for South Africa as a 'JI Intermediary''', *Africa Insight*, no 3

Rowlands, Ian H (1997) 'International Fairness and Justice in Addressing Global Climate Change', *Environmental Politics*, vol 6, no 3, Autumn, pp 1–30

Rowlands, Ian H (1995) *The Politics of Global Atmospheric Change*, Manchester: Manchester University Press

Rowlands, Ian H (1994) 'International Influences on Electricity Supply in Zimbabwe', *Energy Policy*, vol 22, no 2, February, pp 131–43

RSA (1995) *Republic of South Africa Transport Statistics*, Department of Roads of RSA (Republic of South Africa) Pretoria

Røde, Bjørg, Thurlby, Robert and van Zyl, Louis T (1995) 'Why a Southern African Power Pool?', paper presented at the WEC, Regional Energy Forum for Southern and East African Countries, vol 4

SADC (1997) *SADC Energy Sector Action Plan*, Luanda: SADC Energy Sector – TAU, July

SADC (1995a) *Study of the Economics of Natural Gas Utilisation in Southern Africa, Technical Paper B: Southern African Gas Resources and Production Costs*, Luanda: SADC, Technical and Administrative Unit, Energy Sector, May

SADC (1995b) *Study of the Economics of Natural Gas Utilisation in Southern Africa, Technical Paper E: Gas Development Scenarios for Southern Africa*, Luanda: SADC, Technical and Administrative Unit, Energy Sector, May

SADC (1995c) *Study of the Economics of Natural Gas Utilisation in Southern Africa, Technical Paper H: Regional Benefits from Increased Gas Utilisation*, Luanda: SADC, Technical and Administrative Unit, Energy Sector, May

SADC (1990) *Energy Statistics Yearbook*, Luanda: Energy Sector Unit, Technical Unit

SADC–SATCC (1996) *Transport and Communications*, Johannesburg, South Africa, 1–4 February

SADC Treaty (1993) 'Treaty of the Southern African Development Community', reprinted in *International Legal Materials*, vol 32, 1993, pp 116–135

SADC–USA (1998) 'Trade & Investment', http://www.sadc-usa.net/trade/sectors.html; accessed 1 March

SAD–ELEC and MEPC (1996) (Southern African Development through Electricity and Minerals and Energy Policy Centre) *Electricity in Southern Africa: Investment Opportunities in an Emerging Regional Market*, London: Financial Times Energy Publishing

SARDC (1997) 'Trade Protocol: Thorny Road to Ratification', *SADC Today*, Harare: Southern African Research and Documentation Centre, April

Save and Prosper (1998) 'Save & Prosper', http://www.prosper.co uk/prices.htm; accessed 1 March

Scholes, R J and van der Merwe, M R (1996) 'South Africa and the Global Atmosphere' in L Y Shackleton et al (eds) *Global Climate Change and South Africa*, Cleveland: Environmental Scientific Association

Scholes, R J and van der Merwe, M R (1995) *South African Greenhouse Gas Inventory*, Pretoria: Forestek, CSIR

Schweickert, Rainer (1996) 'Regional Integration in Eastern and Southern Africa', *Africa Insight*, vol 26, no 1, pp 48–56

Seabra, Fernando (1993) 'Hydropower, the Forgotten Renewable' in *Hydropower, Energy and the Environment: Options for Increasing Output and Enhancing Benefits, Conference Proceedings*, conference in Sweden, June 1993; Paris: International Energy Agency, pp 45–48

Segal, Aaron (1967) 'The Integration of Developing Countries: Some Thoughts on East Africa and Central America', *Journal of Common Market Studies*, vol 5, no 3, pp 252–82

Shepherd, Anne (1993) 'Building a Bloc', *Africa Report*, vol 38, no 1, January-February, pp 59–63

Snidal, Duncan (1985) 'The Limits of Hegemonic Stability Theory', *International Organization*, vol 39, no 4, autumn, pp 25–57

South Commission (1990) *The Challenge to the South: Report of the South Commission*, Oxford: Oxford University Press

Southern Centre (1997) *FINESSE Activities in the SADC Region: Zimbabwe Country Study*, draft report for the SADC Energy Sector, Technical and Administrative Unit, Luanda

Sperling, D and Deluchi, A (1993) 'Alternative Transportation Energy' in Richard J Gilbert (ed) *The Environment of Oil*, Boston, MA: Kluwer Academic Publishers, Studies in Industrial Organization, vol 17

Söderbaum, Fredrik (1996) *Handbook of Regional Organizations in Africa*, Uppsala: Nordiska Afrikaiinstitutet

Tanesco (1997) personal communication from D E P Ngula, Managing Director, Tanzania Electric Supply Company Limited, Dar es Salaam, 17 January

Thompson, Carol B (1992) 'African Initiatives for Development: The Practice of Regional Economic Cooperation in Southern Africa', *Journal of International Affairs*, vol 46, no 1, Summer, pp 125–144

Tjønneland, Elling Njål (1992) *Southern Africa after Apartheid: The End of Apartheid, Future Regional Cooperation and Foreign Aid*, Fantoft: Chr Michelsen Institute, Department of Social Science and Development

Tsie, Balefi (1996) 'States and Markets in the Southern African Development Community (SADC): Beyond the Neo-liberal Paradigm', *Journal of Southern African Studies*, vol 22, no 1, March, pp 75–98

UK National Communication (1994) 'Executive Summary of the National Communication of the United Kingdom submitted under Articles 4 and 12 of the United Nations Framework Convention on Climate Change', A/AC.237/NC/1, 4 October

UN (1994) *Trends in Environmental Impact Assessment of Energy Projects*, New York: United Nations, Department for Development Support and Management Services, Energy Branch

UNCTAD (1997) *World Investment Directory on Africa*, New York: United Nations, UNCTAD

UNCTAD (1995) *Foreign Direct Investment in Africa*, New York: United Nations, UNCTAD, Division on Transnational Corporations and Investment

UNDP (1997) *Human Development Report 1997*, Oxford: Oxford University Press

UNDP (1996) *Human Development Report 1996*, Oxford: Oxford University Press

UNEP (forthcoming) *Final Report On SADC Regional Mitigation Study*, Roskilde: UNEP Collaborating Centre on Energy and Environment

UNEP (1997) *The Economics of Greenhouse Gas Limitation: Guidelines*, Roskilde: UNEP Collaborating Centre on Energy and Environment, Document 04408 02/02, draft, March

UNEP (1995) *Climate Change Mitigation in Southern Africa: Methodological Development, Regional Implementation Aspects, National Mitigation Analysis and Institutional Capacity Building in Botswana, Tanzania, Zambia and Zimbabwe*, Roskilde: UNEP Collaborating Centre on Energy and Environment

UNEP (1993) *UNEP Greenhouse Gas Abatement Costing Studies: Zimbabwe Country Study, Phase Two*, Roskilde: UNEP Collaborating Centre on Energy and Environment

UNEP (1992) *UNEP Greenhouse Gas Abatement Costing Studies, Phase One Report*, Roskilde: UNEP Collaborating Centre on Energy and Environment, August

UNFCCC (1992) 'United Nations Framework Convention on Climate Change', reprinted in *International Legal Materials*, vol 31, 1992, pp 849–73

Usher, Ann Danaiya (ed) (1997) *Dams As Aid: A Political Anatomy of Nordic Development Thinking*, London: Routledge

Vaitsos, Constantine V (1978) 'Crisis in Regional Economic Cooperation (Integration) Among Developing Countries: A Survey', *World Development*, vol 6, no 6, June, pp 719–769

Vale, Peter (1996) 'Regional Security in Southern Africa', *Alternatives*, vol 21, pp 363–391

Vale, Peter (1982) 'Prospects for Transplanting European Models of Regional Integration to Southern Africa', *The South African Journal of Political Science*, vol 9, no 2

Vale, Peter and Matlosa, Khabele (1996) *Beyond and Below: The Future of the Nation-State in Southern Africa*, Centre for Southern African Studies, no 53

van Horen, Clive (1996) *Counting the Social Costs: Electricity and Externalities in South Africa*, Cape Town: Élan Press and UCT Press

Venter, Denis (1996) 'Regional Security in Southern Africa in the Post-Cold War Era' in Edmond J Keller and Donald Rothchild (eds) *Africa in the New International Order: Rethinking State Sovereignty and Regional Security*, London: Lynne Rienner Publishers, pp 134–148

Watson, Robert T, Zinyowera, Marufu C and Moss, Richard H (eds) (1997) *Summary for Policymakers, The Regional Impacts of Climate Change: An Assessment of Vulnerability*, a Special Report of IPCC Working Group II, published for the Intergovernmental Panel on Climate Change, http://www.ipcc.ch; November; accessed 1 March 1998

WEC (1995a) *Global Transport Sector Energy Demand Towards 2020*, London: World Energy Council

WEC (1995b) *Local and Regional Energy-Related Environmental Issues*, London: World Energy Council, September

Weeks, John (1996) 'Regional Cooperation and Southern African Development', *Journal of Southern African Studies*, vol 22, no 11, March, pp 99–117

World Bank (1995) *Improving African Transport Corridors*, Washington, DC: The World Bank, Operations Evaluation Department, OED Précis, February

World Bank (1996a) *World Development Report 1996*, Oxford: Oxford University Press

World Bank (1996b) *Toward Environmentally Sustainable Development in Sub-Saharan Africa: A World Bank Agenda*, Washington, DC: The World Bank, November

World Bank IENOG (1996) *International Gas Roundtable*, http://www.worldbank.org/html/fpd/ienog/gasround.html; accessed 1 March 1998

World Economic Forum (1997) *Southern Africa: A New Growth Opportunity*, Geneva: World Economic Forum

World Resources Institute (1996) *World Resources 1996–97*, Oxford: Oxford University Press

Young, Oran R (1989) *International Cooperation: Building Regimes for Natural Resources and the Environment*, London: Cornell University Press

'Zambia Needs More Time' (1993) *Southern African Economist*, pp 10–11

ZESA (1997) personal communication

ZESCO (1997) personal communication

ZESCO (1994) *Annual Report 1993/94*, Lusaka: Zambia Electricity Supply Corporation Limited

Zhou, P (1997) *Development of an Efficient Transport Sector in East and Southern Africa: Opportunities and Policy Implications under the UNFCCC*, Nairobi: Energy and Climate Theme Group, African Energy Policy Research Network (AFREPREN) Report

'Zimbabwe – Oil & Gas Industry Overview' (1998), http://www.mbendi.co.za/cyzioi.htm; accessed 1 March

Zormelo, Douglas (1995) *Regional Integration in Latin America: Is Mercosur a New Approach?*, London: Overseas Development Institute, Working Paper no 84, December

Index

Page numbers in **bold** refer to figures, tables and boxes. Those in *italics* refer to notes.

For Product Safety Concerns and Information please contact our EU
representative GPSR@taylorandfrancis.com
Taylor & Francis Verlag GmbH, Kaufingerstraße 24, 80331 München, Germany

www.ingramcontent.com/pod-product-compliance
Lightning Source LLC
Chambersburg PA
CBHW050441280326
41932CB00013BA/2195